The Quests of Leo Patterson: Science, Slime, and Endless Curiosity

Selva Ganapathy Kumaravel

**Universal Declaration of Human Rights (UDHR):
Article 27(1) of the UDHR states**

"Everyone has the right freely to participate in the cultural life of the community, to enjoy the arts and to share in scientific advancement and its benefits."

**International Covenant on Economic, Social and Cultural Rights (ICESCR):
The ICESCR, in Article 15(1)(b), recognizes**

"The right of everyone to enjoy the benefits of scientific progress and its applications."

TABLE OF CONTENTS

A Kid with a Question   1

Robotics Dreams and Freshman Schemes   5

A New Spark – Sophomore Science Explorations   11

Engineering Dreams and Experimental Mishaps   17

The Final Year – Engineering, Explosions, and
Unexpected Victories   23

The Science of Taking a Break   29

Oxford Adventures and the Science of Materials   35

The Sophomore Leap – A Year of Steel, Polymers, and
Puns   41

Third-Year Breakthroughs, Internships, and Laboratory
Mishaps   46

Love, Laughter, and the Great Expedition   55

Research, Romance, and Redefining Science   59

From Conference Applause to Life's Big Pause   64

Settling into the Spotlight   69

The Global Stage and Humbling Returns   74

Lab of Infinite Surprises   79

Renaissance Fair of Science   83

Great Miscalculation   88

Aerogel Adventure   93

Elemental Challenge   98

Liquid Crystal Experiment   103

The Quantum Comedy   108

Piezoelectric Predicament   113

The Misbehaving Magnet Experiment   119

Antigravity Shoes Fiasco   124

Invisible Ink Incident   130

The Quantum Cat Paradox   135

The Molecular Gastronomy Mishap   139

The Great 3D Printer Fiasco   144

The Levitation Experiment Gone Wild   148

The Solar-Powered S'mores Maker   152

The Coffee Catalyst Experiment   157

The Grand Summer of Experiments   162

A Journey Culminating in Stardust   168

The Nobel Surprise and the Next Chapter   173

A Kid with a Question

Leo Patterson had three constants in his life: his blue bike with one wheel slightly larger than the other, his little sister Mia who was obsessed with fossils, and the fact that his favourite subject was science. Leo wasn't exactly sure which branch of science he loved best. One day it was chemistry because he'd seen a video where a guy made things explode with baking soda and vinegar. The next day, it was physics because he discovered that jumping off his couch at exactly the right angle allowed him to flip (almost) without crashing.

And even though Leo's grades in science were sometimes just okay, he always raised his hand in class, ready with questions that made his teacher, Mr. Abernathy, sometimes look at him like he was a creature from a very strange planet.

"Mr. Abernathy, if black holes eat light, are they like cosmic Pac-Men?" Leo once asked, causing the class to erupt in laughter.

Scene Development: Leo's Obsession with Slime - Leo's first experiment happened on a rainy Sunday afternoon in his kitchen. He'd learned that all you needed to make slime was glue and borax, a substance that sounded mysterious and slightly magical.

With a bit of online research, Leo wrote his first scientific hypothesis in a crayon-stained notebook he labelled "LEO'S SCIENCE BOOK":

*"**Hypothesis:** When glue meets borax, it turns into slime. It will be green, stretchy, and maybe glow in the dark. (Leo hadn't confirmed the glow part but was hoping for it.)"*

He got to work. The kitchen quickly turned into a lab: measuring cups, spoons, and various containers were spread across the counter. Leo carefully mixed his ingredients, and, predictably, created a mess of gluey goo that stuck to everything but his hypothesis.

But for Leo, it wasn't a failure. In fact, he was ecstatic. "Mom! It worked! Sort of!" he shouted, holding up a slimy blob that dripped from his hands like some alien life form.

His mother took one look at the mess and sighed. "As long as you're cleaning that up, Leo," she said.

Leo Meets Physics - The following week, Leo's interest pivoted toward physics, which he'd decided was basically just a fancy way to study stuff moving or not moving-and he was going to master it. He began with basic concepts like gravity, friction, and inertia. Leo even attempted to explain these principles to his pet goldfish, Einstein, whom he was convinced was secretly judging his scientific skills.

He'd found out about Newton's Laws of Motion in school, and that evening he tested them out by building a "scientific projectile launcher." It was actually his slingshot, loaded with marshmallows and aimed at a pillow that he pretended was the moon. He loved telling his sister Mia that every time she jumped, she was demonstrating gravity. She didn't find this very interesting until he convinced her that someday she might jump so high she'd reach Mars, at which point she attempted to defy gravity with a series of ambitious leaps around the living room.

Leo's New Quest: Rocket Science - One day, Leo came across a website with instructions for building a "bottle rocket." It promised that by filling a plastic bottle with vinegar and adding baking soda, he could make a rocket launch right from his backyard.

"Mom! I need vinegar! And, uh, baking soda. Like, a lot of it!"

"Leo, if this is another slime experiment…" his mother began, but he was already out the door with a plastic bottle clutched in his hand.

The experiment didn't go exactly as planned. He overfilled the bottle, forgot to aim it away from himself, and was blasted with a face-full of fizzing vinegar. He sputtered and blinked through the mess, feeling a bit like a failed mad scientist.

Still, his notebook recorded it as a success: "Rocket launched. Observed important principle: aim rocket AWAY from self."

Expanding the Story - From here, Leo's experiments and questions keep expanding. He starts exploring **biology** after a class field trip to the zoo, leading him to think he might be able to communicate with animals if he studies hard enough. He learns about **DNA** and becomes convinced he can create a plant that grows marshmallows if he just learns enough about genetics. These attempts lead to funny failures and small triumphs.

Some concepts Leo likes to tackle

Electricity: Trying to generate enough electricity to charge his phone using only lemons, leading to a sticky disaster.

Astronomy: *Attempting to build a telescope out of magnifying glasses and finding a surprise in his own backyard.*

Geology: *Deciding that the "cool rocks" he finds are meteorites and trying to sell them to the local science museum.*

Chemistry: *Attempting to make "homemade rocket fuel," which becomes an elaborate kitchen escapade involving pasta, red pepper, and absolutely no explosions.*

The Never-Ending Journey - Instead of reaching a specific endpoint, Leo's journey is one of endless curiosity. He might not be the smartest kid in the class, but he's definitely the most persistent. His average IQ doesn't stop him from asking questions, diving headfirst into experiments, and getting lost in the wonders of science. And because he keeps learning, failing, and trying again, his story could go on indefinitely.

As Leo grows up, his experiments get slightly more sophisticated (and hopefully, a little safer), but he never loses the quirky enthusiasm that makes him such a lovable budding scientist. He keeps adding entries to "LEO'S SCIENCE BOOK," and by the end, he dreams of adding "Ph.D." next to his name one day-even if he hasn't decided what in yet.

And maybe he never decides; maybe that's the beauty of it. Leo's curiosity is boundless, and so is his story.

Robotics Dreams and Freshman Schemes

Leo Patterson couldn't believe he was finally a freshman. High school felt like a kingdom of opportunity, especially with all the extracurricular clubs. The chess team, the debate club, the drama crew, and, most exciting of all, the Robotics Club. He'd spent most of middle school daydreaming about robotics, a subject that had captured his attention ever since he saw a video of a robot that could beat humans at chess.

"I'm going to build a robot that can mow the lawn and do the dishes," Leo announced to his best friend, Jay, on the first day of school. Jay, who was tagging along to the club meeting with Leo purely out of loyalty, raised an eyebrow.

"Sure, Leo. Just start with a robot that can tie your shoes without setting them on fire."

Leo ignored him. This year, he'd learn to build a real, moving robot.

Scene: Joining the Robotics Club - The Robotics Club was housed in a lab room with half-built contraptions scattered everywhere. Circuit boards, metal frames, wires, and plastic gears covered the tables, while a half-dismantled robot arm sat ominously in the corner. Leo's eyes sparkled; it was like he'd walked into a robotic playground.

The club president, a junior named Lila who looked as though she lived for this kind of thing, welcomed everyone. "Our goal this year," she began, "is to compete in the annual Robotics

Championship. We'll be designing a robot that can navigate a maze, pick up objects, and interact with a simple computer interface."

Leo practically bounced out of his seat. A robot that could interact with a computer? This was way cooler than his middle school experiments with baking soda and vinegar rockets.

"Hey, newbie," Lila called to Leo, who had somehow managed to trip on his way up to the table. "You have any experience with robotics?"

Leo straightened up, determined to look as serious as possible. "I've watched hours of videos on robotics. And I've... made things move."

Jay snorted beside him, muttering, "Pretty sure that thing you made 'move' was just your pet hamster in a plastic ball."

Lila sighed but handed Leo a simple kit to start. It included a breadboard, some resistors, a few wires, and a couple of motors. "Think you can make this run?"

"Absolutely," Leo said confidently, although he didn't really know where to begin.

Leo's First Robot: The Rolling Disaster - In his bedroom that evening, Leo laid out the kit and reviewed the basics of robotics that he'd read in "Robotics for Total Beginners." He was learning about how circuits worked, realizing that electrical engineering was part of the puzzle. Motors, he read, could convert electrical energy into mechanical movement, but getting them to respond to a control system would require a bit of programming, too.

He began wiring the motor to the breadboard, remembering that he'd need resistors to prevent the current from frying the circuit.

"Resistors slow down the flow of electrons," he mumbled to himself. "And that stops things from... catching fire?"

After hours of trial and error, he finally got his motor to whir to life, making a tiny wheel spin. It was just a single wheel attached to a motor on a small platform, but to Leo, it was his first robot. He named it *Wheelie*.

Testing it out on the kitchen floor, he realized two things: one, *Wheelie* only knew how to go in circles; and two, it couldn't stop. The little motorized wheel spun around wildly until it veered off and smacked into the family's refrigerator.

"Mom! Look, I made a robot!" Leo said excitedly.

"Leo, it's stuck under the fridge," she replied, peering down at the rogue creation with an amused smile.

Learning to Code: Robot Reboot - The next Robotics Club meeting was a coding workshop. Leo learned that robots operated by following code, a set of instructions that told the robot what to do. Lila introduced them to Python, a beginner-friendly programming language, and explained how to code movement instructions.

"Think of code as the language you use to talk to your robot," Lila explained. "If you want it to move forward, you tell it to 'move forward.' If you want it to stop, you tell it 'stop.' But don't forget, robots aren't smart. You have to tell them every little thing."

For Leo, coding was a revelation. He was learning that robots were more like very literal toddlers than geniuses-they only did exactly what they were told. He tried inputting simple instructions to control *Wheelie*, telling it to move forward, turn left, and stop at a certain point.

The first time he tried to make *Wheelie* move across the room, it performed an unintentional pirouette and nearly threw itself down the stairs. But by the end of the meeting, he'd managed to get *Wheelie* to follow a simple path on the lab table.

"That's progress, Patterson," Lila said with a nod.

Robotics Club Project: Operation "FetchBot" - As the semester went on, the club began building their competition robot. They'd named it "FetchBot" and designed it to pick up small items in a designated area. Lila explained how sensors worked, teaching Leo and the other freshmen about infrared sensors that could detect obstacles and ultrasonic sensors that could "see" objects using sound waves.

"The sensor emits ultrasonic waves, which bounce off objects and return to the sensor," Lila said. "By measuring the time it takes for the sound to return, the sensor knows how far away the object is."

It blew Leo's mind. He realized that this was similar to echolocation, a natural form of sensing distance that dolphins and bats used. "So, FetchBot is like a bat with wheels?" he asked.

"Basically," Lila laughed. "Only our 'bat' is going to be built with two ultrasonic sensors and the battery life of a tired hamster."

Leo was given the job of assembling the sensors onto FetchBot. With his shaky but persistent hands, he tightened screws, soldered tiny wires, and carefully attached the sensors to

FetchBot's "face." For the first time, he felt like a real robotics engineer.

A Friendly Competition - Near the end of the semester, the club hosted a friendly in-house competition: they split into teams to design small robots that could navigate a maze. Leo, Jay, and a few others were on "Team Circuit Storm." They coded a simple program that instructed their robot to "look" left and right and go forward when there was no obstacle.

But Leo had a crazy idea. He'd read about artificial intelligence and decided he wanted to make his robot "learn" the maze. With some guidance from Lila, he coded a basic memory function, so the robot would record every turn it made and remember which paths led to dead ends.

"Are you sure this'll work, Leo?" Jay asked.

"Maybe. If it doesn't, it'll just wander around confused. Which is sort of like us on a Monday morning."

The robot entered the maze, stopped, turned left, then right, then left again, eventually zig-zagging through the course. It wasn't exactly fast, but it managed to finish the maze. Leo was ecstatic.

Graduation from Freshman Year - By the end of the school year, Leo had come a long way from the kid who couldn't even make a wheel roll in a straight line. FetchBot was ready for the Robotics Championship, and Leo had gained new skills in coding, circuitry, and teamwork.

During the final Robotics Club meeting, Lila presented Leo with a little robot pin, the club's way of welcoming him as an official "roboticist in training." Leo wore it proudly, his head buzzing with ideas for what he wanted to build next year.

On the last day of freshman year, Leo reflected on everything he'd learned. He still had a long way to go before he'd build the kind of robots he'd seen in movies, but he was hooked. Robotics was his calling, and he knew one thing for sure: someday, he'd be a robotics scientist.

And as he stashed away *Wheelie*, his first "robot," he chuckled. Freshman year was just the beginning. Leo's dreams were only getting started.

A New Spark – Sophomore Science Explorations

Sophomore year crept up on Leo faster than he'd expected. He was back at school, ready to dive even deeper into the Robotics Club. But as he settled into his classes and picked up a new schedule, something strange happened: he started noticing all the other sciences he'd never really paid attention to before.

There was biology, with its talk of genes and cells. Chemistry had him looking at the world through the lens of atoms and molecules. And earth science? He'd never thought he'd care about rocks, but suddenly, the idea of understanding Earth's layers seemed... intriguing.

By mid-semester, Leo was feeling restless. Robotics was still amazing, but he was wondering if there might be other fields worth exploring. He didn't want to abandon his robotics projects, but something deep inside him told him to experiment. After all, some of the best scientists had broad interests, right?

Scene: A Flash of Curiosity in Chemistry Class - In chemistry class, they were learning about chemical reactions. Leo's teacher, Ms. Ramos, was explaining the law of conservation of mass, which said that in a chemical reaction, matter is neither created nor destroyed-it's only transformed.

As Ms. Ramos held up a magnesium strip and lit it on fire, Leo watched in awe as it burned with a bright, white glow, leaving behind a white powdery substance called magnesium oxide. "Notice how it burns, but nothing is lost," she explained. "The atoms in the magnesium are simply rearranging to form a new substance with oxygen."

Leo's mind buzzed. This wasn't so different from robotics, where circuits and wires connected to create something new. Here, atoms were like building blocks, sticking together to create entirely new materials.

As Ms. Ramos talked about different types of chemical reactions-synthesis, decomposition, and combustion-Leo couldn't stop thinking about all the possible combinations he could make. What if he could design materials that his robots could use? Heat-resistant alloys, or maybe even light-sensitive coatings?

It was like a door had cracked open. Chemistry wasn't just some abstract subject; it was a toolkit for making things in the physical world.

Scene: Discovering DNA in Biology - Around the same time, Leo's biology class began a unit on genetics, exploring how DNA is like a blueprint for living organisms. His teacher, Mr. Chen, explained how DNA is composed of just four nitrogenous bases-adenine (A), thymine (T), cytosine (C), and guanine (G)-which pair up to form the rungs of the DNA ladder.

"This," Mr. Chen said, showing a double helix model, "is the instruction manual for life. All the information needed to create every part of a living thing is encoded here."

Leo was blown away. DNA sounded a lot like the binary code he'd been learning in robotics. In robots, a combination of 1s and 0s could tell a machine what to do, but in biology, a sequence of As, Ts, Cs, and Gs could shape an entire organism. It got him thinking: could he somehow combine his passion for robotics with genetics? Could there be a way to create "bio-robots" that mimicked the natural processes of life?

Suddenly, he was enthralled by the idea of biotechnology, a field that blended biology and engineering.

Experiment: The Algae Project - Eager to try something hands-on, Leo dove into a school science project on biotechnology. He proposed a project on biofuels-specifically, creating fuel from algae. Leo read about how algae, a simple plant that can grow in water, produces oils that could be converted into biofuel. Unlike fossil fuels, algae grow quickly and are renewable, making them a potential solution to sustainable energy.

"Algae might not look cool, but it's got power," Leo told Jay, who was less than impressed. "In the right setup, algae can produce oil and oxygen. It's like a mini green factory!"

To test this, Leo set up a small algae-growing station in the school lab, using a few glass jars, an aquarium pump to aerate the water, and a small LED light to simulate sunlight. He kept a journal of his observations, noting how the algae grew faster with more light and oxygen. He even tested extracting oil from the algae, which involved a complicated process of drying the algae and breaking down its cell walls to release the oil.

After weeks of tinkering, Leo managed to extract a tiny bit of oil. It wasn't enough to power anything, but it was enough to get him thinking about the potential of renewable energy sources. "Who knows," he thought, "maybe I'll be the one to invent algae-powered robots someday!"

An Unexpected Discovery in Earth Science - One rainy afternoon, Leo's earth science teacher, Ms. Keller, gave the class an assignment to research rocks and minerals. Leo wasn't expecting much, but he stumbled across a fascinating mineral called *pyrite*, also known as "fool's gold" for its shiny, gold-like appearance. As he read further, he found out that pyrite wasn't just pretty-it had a practical use in science as well.

Pyrite, he learned, was conductive and even capable of producing a tiny spark if struck correctly. It had been used historically in early firearms to ignite gunpowder.

Inspired, Leo got his hands on a small sample of pyrite for his project and conducted experiments, trying to generate sparks by striking it against other materials. He also discovered that certain rocks, like quartz, generated a small electric charge when squeezed, a property called *piezoelectricity*.

"You could make a battery out of rocks," Leo muttered excitedly, jotting down ideas in his notebook. The gears in his mind were turning. What if there were ways to harness minerals like quartz and pyrite to power small devices? Or what if these materials could be used to power environmental sensors in remote places?

For Leo, earth science was no longer just about old rocks-it was a resource waiting to be tapped for the next generation of energy.

Exploring Astronomy: The Cosmos Beckons - In addition to his new interests in biology and earth science, Leo's curiosity drifted to the stars. His parents got him a small telescope for his birthday, and he'd been spending nights staring at the moon, Jupiter's moons, and, on clear nights, the hazy band of the Milky Way.

Leo read about how some stars are millions or even billions of light-years away. Each time he looked at the stars, he felt like he was looking back in time. The light from these stars had travelled across the universe, sometimes for eons, before reaching his eyes.

"A star's light might take thousands of years to get here," he explained to Jay one afternoon. "So when you look at a star, you're actually seeing it as it was thousands of years ago."

Jay tried to keep up, nodding slowly. "So... it's like a space version of a delayed text message?"

Leo laughed. "Exactly! And some of those stars might not even exist anymore. We're just seeing the light that left them long ago."

This idea of cosmic "time travel" fuelled his fascination with astronomy, and he began considering the possibility of one day studying astrophysics, imagining himself working on a team to discover new planets or explore the nature of black holes.

Sophomore Year's End: A Budding Scientist with Many Dreams - By the time finals rolled around, Leo's passion for robotics had only grown, but now it was joined by interests in chemistry, biology, geology, and astrophysics. He hadn't abandoned his dreams of robotics; he still attended Robotics Club meetings and continued to work on FetchBot 2.0 with his team. But now, he felt like there was an entire universe of scientific possibilities out there, waiting for him to explore.

On the last day of school, Leo packed up his locker with a strange sense of satisfaction. Sophomore year had been about discovery-not just of new science topics, but of new interests and dreams. He realized that science wasn't a single path but rather a web of interconnected fields, each one adding a new layer to his understanding of the world.

That summer, Leo kept experimenting, combining all his newfound knowledge. He dreamed of building a robot that could harness algae for fuel, generate electricity from minerals, and even use sensors inspired by biological systems.

As he closed his journal and stashed away his chemistry notes, Leo smiled. Sophomore year had expanded his world, and he knew that he'd carry these new interests with him into junior year. For Leo, the journey was far from over; in fact, he was just getting started.

Engineering Dreams and Experimental Mishaps

Junior year hit Leo Patterson like a hurricane. As he walked through the halls on his first day back, he was overwhelmed by the sense that high school was moving fast, and he'd soon have to decide on a career. Over the past two years, he'd dipped his toes into biology, chemistry, and even astronomy, but the one subject that kept calling to him was engineering. Engineers were the ones who built things, from bridges and rockets to robots and renewable energy devices. Leo felt a rush of excitement just thinking about it. He'd be an inventor-a maker.

But before he could fully dive into his engineering dreams, he was drawn to a new club that had started up that year: the School Science Club. Run by Mrs. Vega, an enthusiastic physics teacher, the club offered hands-on experiments, field trips, and, best of all, a chance to compete in the annual Small Science Competition.

Scene: The Science Club and Leo's Big Idea - The School Science Club was a gathering of science-loving students from all grades, and the meetings were a mix of creative chaos and science demos. Leo joined his first meeting with his usual enthusiasm, sitting down among beakers, test tubes, and random bits of equipment. Mrs. Vega introduced herself with a booming voice.

"This year, our main goal is to enter the Small Science Competition. You'll each create a project or experiment, and if yours is chosen, you'll get to present it on stage at the competition in front of everyone!" she announced.

Leo was buzzing with ideas. He wanted to create something engineering-inspired, something fun and, ideally, explosive (but not in a dangerous way). After brainstorming with his clubmates, he decided to make a machine that could launch small rubber balls accurately using a homemade catapult.

"Picture it!" he told Jay one afternoon, showing him a hastily drawn diagram. "A rubber band-powered catapult that'll hurl a rubber ball with pinpoint accuracy! It'll be like medieval engineering meets physics."

Jay squinted at the diagram. "So, you're building a rubber band slingshot for... what, ping-pong balls?"

"Not just any slingshot," Leo replied with a wink. "This one's going to be accurate, efficient, and scientifically advanced. It'll have angles, measurements, and-okay, maybe a bit of luck, too."

Jay looked sceptical but supportive. "Just try not to launch it into anyone's face."

Leo's Engineering Journey: Learning About Trajectory and Tension - As Leo began building his catapult, he dived into the science of trajectory, tension, and kinetic energy. He read about *elastic potential energy*, the energy stored in a stretched or compressed elastic material, like a rubber band.

"Basically," he explained to Mrs. Vega during a club meeting, "the rubber band stores energy when I stretch it. When I release it, the energy converts to kinetic energy and launches the ball!"

"Impressive, Leo," Mrs. Vega said, smiling. "But don't forget to control your angle of launch-aim too high, and gravity pulls it back down too soon; aim too low, and it doesn't go far enough."

Leo spent hours testing different angles, learning how the angle affected the distance his rubber balls would travel. He realized that the ideal launch angle was around 45 degrees for the farthest distance. The next challenge was accuracy, which was tougher to perfect, given the inconsistency of his homemade setup.

After a lot of tinkering, Leo finally had a working prototype. The catapult was a crude but endearing structure, complete with a rubber band launch system, a wooden arm, and a cardboard "sight" for aiming. It looked like something from a science-fair thrift shop, but it worked... mostly.

The Small Science Competition: Leo's Catapult Takes the Stage - The day of the Small Science Competition arrived. Leo's heart pounded as he set up his catapult on stage, surrounded by an audience of fellow students, parents, and teachers. Jay was in the crowd, trying not to laugh at the sight of Leo's humble contraption. Next to him, other students displayed polished experiments, from volcano models and circuit boards to chemical reactions with eye-catching colours. Leo felt his face flush, but he took a deep breath.

"It's not about looking impressive; it's about the science," he reminded himself.

As it was his turn, he stepped up to the microphone. "Hi everyone! I'm Leo Patterson, and this is my catapult experiment. I've designed a rubber band-powered launcher that uses elastic potential energy to propel a rubber ball with precision and accuracy... I hope."

The audience chuckled, and Leo felt his nerves settle a bit. He loaded a rubber ball into the catapult, aimed carefully, and pulled back the rubber band.

"For science!" he declared as he released the arm.

The ball shot forward-straight at Mrs. Vega's table. With a loud *splat*, it landed squarely in a bowl of punch, splashing red liquid everywhere.

The crowd erupted in laughter. Mrs. Vega looked surprised but tried to hold back her smile. Leo's face turned bright red, but he couldn't help grinning.

"Well," he said, "I guess that was my unplanned 'splash zone' experiment. Science has a way of surprising us, right?"

The audience laughed and clapped, entertained by the spectacle. His catapult might not have been the most accurate device in the competition, but it was definitely the most memorable.

A Lesson in Engineering: Analysing the "Splash Incident" - After the competition, Leo sat down with Mrs. Vega, who offered to help him analyse what had gone wrong. Together, they dissected the "splash incident," reviewing the physics behind it.

Mrs. Vega pointed out, "Your catapult was great in terms of power, but remember that without a consistent aiming mechanism, you'll always get a bit of randomness in the launch. Engineering is about designing for precision and control."

"Precision and control," Leo repeated, nodding. He realized that his catapult had worked, but it had lacked the precision he'd needed to make it truly effective. If he'd added a guide rail or a weight-balancing mechanism, he might have been able to predict where the ball would land every time.

But even in failure, Leo learned a valuable lesson: in engineering, every "mistake" was simply data, feedback that pointed toward improvement.

Exploring New Passions in Physics and Engineering - Inspired by the competition and his talk with Mrs. Vega, Leo continued his journey in engineering. He began reading about mechanical engineering, learning how different materials could affect an invention's performance. He experimented with building small bridges from Popsicle sticks to learn about *tensile strength* (how much a material can stretch before breaking) and *compressive strength* (how much a material can be compressed before failing). It was eye-opening to see how physics and engineering principles worked in tandem to build everything around him, from skyscrapers to smartphones.

One evening, he discovered electromagnetism in his physics textbook. Leo learned that electric currents flowing through wires created magnetic fields. By wrapping a wire around a nail and connecting it to a battery, he could create an electromagnet strong enough to pick up small objects like paper clips.

"An engineer has to know how to control electricity, too," he told Jay as they watched his homemade electromagnet pick up nails from the workbench.

"Yeah, but don't you think that's like... magic?" Jay asked, still baffled by the concept of invisible magnetic fields. "You're making magnets out of thin air!"

Leo grinned. "Science is the closest thing we've got to magic, right?"

End of Junior Year: A Growing Engineer and Scientist - By the time junior year was winding down, Leo had learned so much more about engineering and the science that underpinned it. He'd tried-and failed-more experiments than he could count, but each one had taught him something valuable. He'd built more rubber-band contraptions, created electromagnets, and even dabbled in coding again to program a small motor for a project. He still wasn't sure what kind of engineer he wanted to be, but he knew one thing for sure: engineering was the field for him.

On the last day of school, he walked down the hallway with a proud grin, reflecting on his year. He might not have taken home a trophy at the science competition, but he'd gained something far more valuable: the thrill of experimentation, the joy of building, and the knowledge that even the biggest failures could make for the best memories.

And as he packed away his catapult and science journals, Leo smiled, knowing that his senior year would be his best yet.

The Final Year – Engineering, Explosions, and Unexpected Victories

Leo Patterson couldn't believe he was a senior. This was it: his last year of high school. College applications loomed, but Leo had his heart set on one goal before he walked across that graduation stage: to make a mark as an engineer. Over the summer, he'd decided that his passion lay in mechanical engineering, the art of building and designing machines. To him, it was the perfect blend of science, problem-solving, and creativity.

Leo's senior year began with a new adventure. The School Science Club was planning a big project: the team would compete in an international science competition. The catch? It would be held halfway across the world in Geneva, Switzerland. Leo's heart raced at the idea. Competing on an international stage sounded epic, but it also meant he had to create an experiment that would be both impressive and reliable. He started brainstorming right away, jotting down ideas on everything from kinetic energy projects to a weird, half-baked idea involving potato batteries.

The Invention: The "Dancing Bot" - In the end, Leo decided on a project he called the "Dancing Bot," a small robot designed to dance based on vibrations generated by sound waves. The bot had a motor attached to a mini platform with springs that could vibrate in response to music or sound. Using sensors, the bot would "dance" in sync with any song, its movements modulated by the different frequencies of sound. The science behind it was rooted in resonance-how certain frequencies could make objects vibrate more intensely- and Leo thought it would be a crowd-pleaser.

Leo spent weeks testing it, tweaking the springs, adjusting the sensors, and making sure the bot could react to various sounds. He even enlisted his friend Jay to test it out with different music genres.

"Okay, let's try classical music," Leo instructed, setting the bot on his workbench.

Jay played Beethoven's "Symphony No. 5," and the bot moved in slow, calculated movements, almost as if it were contemplating the music.

"Not bad!" Jay said, laughing. "Let's see how it handles some dubstep!"

As soon as the heavy bass dropped, the bot went into overdrive, practically launching itself off the bench. It landed with a clatter, one wheel still spinning.

"Guess it's not a fan of dubstep," Leo joked, scribbling notes in his engineering journal.

Preparing for Geneva: The Science Club Rallies Around Leo - As the date for the international competition neared, Leo's School Science Club rallied around him. Mrs. Vega, the club sponsor, even organized mini "concerts" during lunch so that Leo could test his bot in front of a live audience. The tests went... mostly well. There were a few near catastrophes, like the time the bot decided to tango with Mrs. Vega's coffee cup, but each mishap only made Leo laugh harder. He embraced each little failure, learning how to better stabilize his bot with every wobble and bounce.

His goal for Geneva was to present a demonstration that was scientific, entertaining, and, ideally, not dangerous. But as the competition date drew closer, Leo couldn't shake his nerves. He'd seen videos of past entries-students presenting elaborate genetic engineering projects, complex chemical reactions, and even mini wind turbines. And here he was with a dancing robot that occasionally did its own thing.

Scene: The International Science Competition in Geneva - The day of the competition in Geneva was a whirlwind. Leo walked into a huge conference hall filled with young scientists from all over the world. The atmosphere buzzed with energy, and students wore lab coats, goggles, and gloves. Some projects had intimidating setups: laser experiments, holographic displays, even a 3D-printed model of a rocket.

As Leo set up his Dancing Bot, he felt a flash of doubt. Would his quirky project stand a chance against such advanced experiments?

When his turn finally came, he took a deep breath and stepped onto the stage, facing a panel of judges and a large audience. He started with his introduction.

"Hi! I'm Leo Patterson from the School Science Club in Maplewood High, and today I'm presenting the 'Dancing Bot,' a robot that can dance in response to sound waves."

He looked out into the crowd, where he spotted Mrs. Vega giving him a thumbs-up. He pressed play on his chosen song, a jazzy tune with a lively beat, and watched as the bot began to sway to the rhythm. The audience chuckled and clapped along.

"So, here's the science part," Leo continued, trying to sound professional. "The bot reacts to sound waves. When the sound hits certain frequencies, it creates vibrations that make the bot bounce on its springs. Different frequencies result in different moves, giving it this... groovy effect."

But as he reached the climax of the song, disaster struck. Someone in the audience sneezed loudly, and the bot, confused by the sound, veered wildly to the right. It spun in circles, did a little hop, and then, in a grand finale, it tipped over dramatically, landing on its back like an exhausted breakdancer.

The audience erupted into laughter and applause. Leo grinned, feeling his face turn red, but he owned it.

"Well, folks," he said, picking up the bot and giving it a pat, "sometimes, science has a mind of its own."

The judges smiled, clearly entertained, and as he left the stage, he could still hear the laughter and cheers behind him. It wasn't exactly the graceful performance he'd planned, but he'd left an impression.

Winning the Competition – In the Most Leo Way Possible - Later that evening, the winners were announced. Leo didn't expect much, but he stood with his clubmates anyway, politely clapping for the participants who took third and second place.

"And first place goes to... Leo Patterson from Maplewood High School for the Dancing Bot!"

Leo's jaw dropped as his friends cheered and shoved him toward the stage. The judges handed him a trophy and explained their decision.

"Leo's project combined engineering principles with creativity and humour, making science approachable and fun," one judge said. "He reminded us that sometimes, the joy of science is in its surprises."

Leo held the trophy high, grinning from ear to ear. In his heart, he knew that winning wasn't about having the most flawless project-it was about inspiring others to see science as something fun and accessible.

Valedictorian Speech with a Science Twist - Back home, Leo's final achievement awaited him: delivering the valedictorian speech at graduation. To everyone's surprise, he'd been chosen for his perseverance, his humour, and his boundless curiosity. He stood on the graduation stage in his cap and gown, looking out at his classmates and teachers.

"Hello, everyone," he started, taking a deep breath. "I know I'm supposed to give an inspiring speech about the future, but I'd like to do it in my own way... with a little science experiment."

The audience leaned in, curious. Leo had brought a small balloon, a bit of baking soda, and a bottle of vinegar on stage. He poured the vinegar into the bottle, quickly added the baking soda, and placed the balloon over the opening. The balloon inflated with carbon dioxide gas, expanding rapidly as the audience laughed and clapped.

"See, this reaction represents life," he said, as the balloon kept expanding. "You add a little bit of yourself, a little bit of new experience, and you never know what you're going to get. It might not always go as planned-trust me; my bot would agree-but sometimes, the best results are the unexpected ones."

With that, he popped the balloon (safely), drawing a last laugh from the crowd.

Leo continued, "In these past four years, I learned that science isn't just a subject. It's a way of looking at the world with curiosity, even when you fail, and especially when things get messy. Here's to all of us making our own discoveries, big or small, and to always having a little fun along the way."

As he finished, the crowd gave him a standing ovation. Leo felt a sense of pride and joy as he looked out at his friends, teachers, and family. It was a moment he'd never forget.

Graduation and New Beginnings - As he walked across the stage to receive his diploma, Leo knew he was ready for whatever came next. His love for science and engineering was stronger than ever, but he knew that this was just the beginning. From robotics to engineering mishaps, every experiment had taught him to embrace the unexpected, to laugh through failures, and to keep pushing forward.

With his diploma in hand and dreams of becoming an engineer, Leo stepped into the next chapter of his life, ready to explore, invent, and maybe make a few more delightful mistakes along the way.

The Science of Taking a Break

The summer before college was Leo Patterson's first real break in years. High school was behind him, and while college loomed on the horizon, for the first time, he had time to breathe. But, of course, Leo wasn't content to just lie on the couch all summer. He wanted to dive deeper into science, explore new fields, and maybe, finally, figure out what he wanted to study.

One sunny afternoon, he found himself sketching ideas in his notebook. He'd filled pages with brainstorms ranging from artificial intelligence to environmental engineering. He felt as if the entire universe of science was open to him. But where to start?

Leo's friend Jay came over one evening, as always, ready to see what new project Leo was working on.

"So… what now?" Jay asked, eyeing the stacks of books piled on Leo's desk. "I thought you were going to relax."

"I am relaxing," Leo grinned. "I just thought it'd be fun to explore some science experiments. See what really sparks my interest, you know?"

Jay chuckled. "Sure, Leo. 'Relaxing' with a chemistry set and a textbook on quantum mechanics. Totally normal summer activities."

Experiment 1: Homemade Spectroscopy with a CD - Leo's first experiment was a homemade spectroscope, a device used to view the light spectrum emitted by different light sources.

The idea was simple: by looking at light through a diffraction grating-in this case, an old CD-he could break it down into its component colours, like a rainbow.

He explained the science to Jay, who looked sceptical but curious. "So, light looks white to us, but it's actually made up of all the colours in the spectrum. A spectroscope splits the light so we can see each colour separately, which is how scientists can figure out what stars are made of, even if they're billions of miles away."

Leo held up his makeshift spectroscope, a CD taped to one end of a cardboard tube. When he aimed it at a lamp, the light split into vibrant bands of red, green, blue, and violet.

"Whoa, it's like a rainbow!" Jay said, peering through it. "So, you could actually tell if a star has hydrogen or helium just by looking at its light?"

"Exactly!" Leo said, proud of his simple invention. "It's all in the colours. Hydrogen gives off specific colours, and helium gives off others. It's like a code we can read."

Jay grinned. "Okay, that's actually pretty cool. What's next?"

Experiment 2: Building a Solar-Powered Charger - After the spectroscope, Leo decided to try his hand at something more practical: a solar-powered phone charger. He had learned that solar energy is harnessed using photovoltaic cells, which convert sunlight into electricity. The science behind it amazed him. Photovoltaic cells were based on the *photoelectric effect*, where light striking a material-typically silicon-causes electrons to flow, creating an electric current.

With some help from his dad, who found a small solar panel online, Leo pieced together a basic charger. He connected the solar panel to a USB port, then to a battery pack. After a few tests, he took it outside on a sunny day.

"Here goes nothing," he said, plugging in his phone. To his amazement, it worked! The battery pack charged up slowly but steadily under the sunlight. He couldn't believe he'd made a working phone charger from a few wires and a solar panel.

"It's official," Leo said to Jay, showing off his newly charged phone. "I've harnessed the power of the sun."

Jay laughed. "I have to admit, that's pretty awesome, Leo. So, are you gonna be an environmental engineer now?"

"Maybe," Leo said, grinning. "But I'm keeping my options open."

A Family BBQ and Leo's Science "Demo" - The Patterson family's annual summer BBQ was the perfect chance for Leo to show off his experiments to his extended family. He decided to bring his solar charger and spectroscope to share what he'd been working on. His younger cousins were fascinated, especially when he demonstrated the spectroscope by showing them the spectrum from the grill flame.

"Fire gives off a spectrum too, depending on the chemicals inside," he explained, making sure his cousin Peter was far enough from the grill. "Different colours mean different elements. For example, sodium makes flames yellow, and copper can make flames green."

Peter looked awestruck. "You're like a science wizard, Leo!"

Leo laughed. "I'm just trying to make science fun."

Experiment 3: The Magnetic Levitation Train Model - Leo's last big project of the summer was inspired by a documentary he'd seen on magnetic levitation trains, or maglevs, which use powerful magnets to float above the track and reduce friction. It sounded like pure sci-fi, but maglev trains were real and could reach speeds over 300 miles per hour.

He decided to make a model to understand the concept. Using a few strong neodymium magnets and some lightweight track he found in the garage, he created a tiny maglev car. When he aligned the magnets on the car and the track just right, the car levitated-barely, but it was hovering.

"Look at that!" he said, watching the car float above the track. "Magnetic fields repel each other, so with enough force, we can actually make something float. That's how maglev trains work-they're basically floating above the track, so there's no friction holding them back."

Jay was astonished. "Okay, Leo, this is your best project yet. You're literally making things float!"

"Science does have a way of defying gravity," Leo replied with a grin.

A Heart-to-Heart with His Parents - One evening, Leo's parents sat down with him to discuss his college plans.

"We're really proud of you, Leo," his mom said, smiling. "But have you thought about what you want to study? It seems like you're interested in everything."

Leo nodded. "That's the problem-I am interested in everything. But I think it'll come down to engineering. There's so much to learn and explore in mechanical engineering, and I could branch out to other fields from there."

His dad gave him a supportive nod. "Sounds like a great plan, but don't be afraid to keep experimenting. Sometimes your passion finds you when you least expect it."

His parents' advice resonated with him. Leo knew he wanted to make things, but he decided to keep an open mind. There was no reason he couldn't dabble in other fields-especially if he had the whole world of science at his fingertips.

Acceptance to the University of Oxford - As the summer drew to a close, Leo's college acceptance letters began arriving. He opened each one with anticipation, feeling the weight of each decision.

One evening, as he returned from an outing with Jay, his mom was waiting by the door, holding an envelope. Leo's heart skipped a beat. The envelope was from the University of Oxford.

With trembling hands, he tore it open and read the letter. His eyes widened, and he felt a grin spreading across his face.

"Mom! Dad! I got in-I got accepted to Oxford!"

His family erupted into cheers, and even Jay joined in, giving him a high-five.

"This is incredible, Leo!" Jay said. "You're going to Oxford! That's like... the big leagues!"

Leo felt an immense sense of pride and excitement. It was the culmination of years of experiments, late-night science readings, and countless inventions. He'd finally achieved his dream, and he was ready to take on whatever challenges lay ahead.

Farewell to Home, and Looking Ahead - As summer faded into fall, Leo packed his bags for Oxford, ready to begin his next adventure. He felt a mix of excitement and nerves, but he knew one thing for sure: he was finally on the path to becoming an engineer.

As he said goodbye to Jay, his family, and his friends, he reflected on his journey. Each experiment, from failed catapults to dancing robots, had taught him something valuable. He knew he'd continue learning, failing, and succeeding, and he couldn't wait to see what the future held.

Leo boarded the plane to Oxford with a heart full of dreams, ready to turn his passion for science into something that could change the world.

Oxford Adventures and the Science of Materials

Leo's first day at Oxford University felt like stepping into a world straight out of a novel. The old stone buildings, cobbled paths, and ivy-covered walls were enchanting, and yet the historic charm of it all made him feel a little... lost. He chuckled to himself-maybe he'd find his way by looking for a "You Are Here" map. Oxford was thrilling, but it was also a bit daunting.

His major? Materials Science. After months of contemplation, Leo had decided that materials science was where he wanted to begin his journey. The field offered everything he was interested in the study of what things are made of, how they work, and how to create new materials that could do things no one had ever imagined.

First Lab: The Wonders of Polymers and an Unexpected Goo Incident - Leo's first lab session in materials science introduced him to polymers, long chains of molecules that make up materials like plastic, rubber, and even the proteins in our bodies. Polymers were essential in everything from creating flexible electronics to designing medical implants.

In the lab, their task was to create a hydrogel-a substance that could absorb water and swell up, useful in everything from diapers to wound dressings. Leo couldn't wait to get started. He mixed a solution of polymer powder and water, watching as it turned into a stretchy, gooey substance.

One of his lab partners, Emily, smirked as she watched Leo carefully lifting his glob of hydrogel. "Looks like you're playing with alien slime."

Leo laughed. "Hey, this is the cutting edge of diaper technology! This little glob could save a parent's sanity someday."

Emily raised an eyebrow. "Just don't drop it."

Naturally, as if on cue, the hydrogel slipped from Leo's fingers and splattered on the table, slowly oozing over the edge and onto the floor.

"Oh no!" Leo groaned, trying to scoop it up with a spatula.

The instructor noticed and laughed. "Hydrogels are sticky business, Leo. Consider it a lesson in materials handling."

The entire lab burst into laughter, and though Leo's face turned red, he laughed along. The class was off to a sticky, but memorable, start.

Discovering the Superpowers of Graphene - As the weeks went on, Leo was introduced to one of the most exciting materials in science: graphene. Graphene is a single layer of carbon atoms arranged in a hexagonal lattice, and it's the strongest, thinnest material known to science. Just a single sheet of it could support over 100 times its own weight!

The lecturer explained, "Graphene conducts electricity better than copper, is flexible, and is nearly transparent. It has applications in electronics, energy storage, and even medicine."

Leo's eyes widened with excitement. He imagined creating ultra-thin solar panels, bendable phone screens, and even super-efficient batteries using this wonder material. For his next project, he decided to try making a small, flexible circuit board out of graphene-based ink.

He sketched out a simple circuit design and, with a small brush, painted the graphene ink onto a flexible plastic film. After letting it dry, he connected a tiny LED light to the circuit and flicked the switch. To his delight, the light blinked on, powered by his makeshift circuit.

One of his new friends, Sam, looked over. "You're telling me that stuff is just one atom thick and can conduct electricity like that?"

Leo nodded. "Yep! It's as close as we've come to science-fiction materials in real life."

Sam grinned. "That's insane, man. Are you sure it's safe? What if it develops a mind of its own?"

Leo laughed. "If this LED starts talking to me, we'll have a whole new problem."

Cultural Adventures and the Great Tea Fiasco - Adapting to life in the UK had its own quirks. For one thing, tea was everywhere. Tea was the go-to drink during lectures, study sessions, and even lab breaks. Leo thought he knew tea-until he tried making a cup in front of his dorm mates.

"Leo," Emily said, eyeing him as he microwaved his mug of water, "you're making tea... in the microwave?"

Leo looked up, confused. "Uh, yeah? That's how we did it back home."

The room fell silent as everyone stared, horrified. Sam's mouth dropped open.

"Mate, this is England," he said solemnly. "You don't microwave the water. You have to boil it in a kettle, pour it over the tea bag, and let it steep."

Leo felt like he'd broken an ancient British code. "Alright, alright! I'll do it the proper way next time."

They all laughed, and from that day on, Leo received daily "tea lessons" from his friends, who were determined to teach him the "correct" way to make a cup of tea.

Experimenting with Alloys and a Near Disaster - One of Leo's favourite classes was on alloys-metals combined with other elements to improve their properties. From stainless steel to bronze, alloys are essential in everything from construction to electronics.

For his final project of the semester, Leo decided to create his own custom alloy, one that combined the strength of titanium with the flexibility of aluminium. His plan was to melt and mix a small sample of each metal, then cast it into a thin sheet that could bend without breaking.

After triple-checking his calculations, Leo set up his mini furnace, donned his safety goggles, and began melting the metals. But as he poured the mixture into a mould, he misjudged the amount, and a blob of molten metal splattered onto the table.

The room went silent as everyone held their breath. Fortunately, the metal cooled quickly without catching fire, and Leo sighed in relief.

"Well, that was almost a real 'metal meltdown,'" he joked, getting a few chuckles from his classmates.

In the end, his alloy didn't turn out exactly as planned, but he learned a lot from the experience. His sheet was a little too brittle, but he was excited to refine the process and try again.

Weekend Adventures and Inventing a Foldable Bike - Outside of class, Leo spent weekends exploring Oxford with his friends. One Saturday, after a long hike, they found themselves talking about their least favourite part of walking around the city: carrying heavy backpacks.

"That's it!" Leo said, snapping his fingers. "What if we had a bike that could fold up small enough to fit in a backpack?"

His friends laughed, but Leo's wheels were already turning. That night, he sketched a design for a foldable bike made from lightweight materials like aluminium and carbon fibre. He imagined a bike frame that could fold into three parts and fit snugly into a specially designed backpack.

The next week, he brought his idea to the workshop. The process took longer than he expected-getting the hinges right was tricky, and finding the balance between strength and weight was a challenge. But after a few test rides and a lot of adjustments, he finally had a working prototype.

On a sunny Sunday, Leo unveiled the bike to his friends. He unfolded it with a flourish, got on, and started pedalling down the street, backpack swinging on his back.

Emily was amazed. "That's incredible, Leo! It actually works!"

Sam chuckled. "Alright, but can it handle my weight?"

Leo laughed. "You're welcome to try it! Just… be gentle on the hinges."

They spent the rest of the day taking turns riding the foldable bike and brainstorming ideas to make it even better. It was moments like these that reminded Leo why he loved engineering: solving problems, innovating, and having fun with friends.

The Move to Second Year - As the end of the year approached, Leo reflected on how far he'd come. His first year at Oxford had been filled with experiments, new friendships, and unforgettable experiences. He'd learned about everything from polymers to graphene, created his own circuit boards, and even invented a foldable bike. Materials science had proven to be the perfect major for him, and he couldn't wait to dive deeper into the world of composites, nanomaterials, and smart technologies.

On the last day of term, Leo and his friends celebrated with a picnic by the River Cherwell. They shared stories from the year, laughed about the tea incident, and made plans for the future.

As he looked around at his friends and the city that had become his second home, Leo felt a profound sense of gratitude. He was excited, a little nervous, and ready for whatever challenges lay ahead in his second year at Oxford.

With new ideas already swirling in his mind, Leo knew that his journey in science was only beginning.

The Sophomore Leap – A Year of Steel, Polymers, and Puns

Leo Patterson's second year at Oxford promised to be bigger and better than the first. He now had a solid group of friends, a deeper understanding of his chosen field-materials science-and a growing reputation among his peers as "the guy who once invented a foldable bike that squeaked like a mouse under pressure."

The early days of his sophomore year were filled with excitement. Leo moved into a new dorm room with Sam and a third roommate, Naveen, who had an encyclopaedic knowledge of random science facts and a dry sense of humour. They all quickly bonded over late-night debates about whether graphene could be the answer to everything, or if that title belonged to duct tape.

The First Major Lab: Metallurgy and the Miracle of Steel- Leo's first big project of the year involved studying steel and its myriad uses in construction and industry. Steel, an alloy primarily of iron and carbon, was fascinating in its simplicity and strength. He marvelled at how adding just a tiny percentage of carbon could transform iron from a brittle lump into a material that held up skyscrapers.

Professor Greenwell, a burly man with a voice like a foghorn, led the lab. "Today, we will be testing the tensile strength of different steel samples," he announced. "Leo, please don't drop the equipment this time."

Leo blushed as the class burst into laughter. He'd become notorious for his hydrogel incident the year before, but he owned it with a sheepish grin. "No promises, Professor."

As they tested samples, Leo discovered how alloys like stainless steel got their corrosion resistance from chromium and nickel. He looked at Sam, who was struggling with the grips on the testing machine. "You know," Leo said with a smirk, "I heard if you whisper 'You're strong and independent' to the steel, it performs better."

Sam rolled his eyes. "Oh, really? Did that work on your foldable bike?"

They both laughed, drawing another amused glance from Professor Greenwell. This was the magic of materials science: learning serious, world-changing facts while slipping in a few jokes.

Naveen's Obsession and the Polypropylene Debacle - Living with Naveen meant that random science facts popped up at unexpected moments. One evening, as Leo and Sam were studying polymers in their dorm, Naveen burst in, eyes wide.

"Did you know that polypropylene is in almost every plastic thing we touch?" he said, holding up a plastic spoon as if it were a prize. "You're practically studying a celebrity!"

Leo looked up from his notes. "Good to know, but if that spoon starts singing, I'm moving out."

The trio decided to make a project out of testing the properties of everyday polypropylene items. Leo constructed a makeshift tensile strength machine using an old exercise resistance band and a scale he borrowed from Emily, who raised an eyebrow when he told her why he needed it.

Their first test was a plastic fork. As Leo slowly applied pressure, Sam took notes, and Naveen narrated like a nature documentarian. "Behold, the wild polypropylene fork, known for its remarkable resistance to breaking under stress until-"

Snap! The fork broke and pinged off the wall, narrowly missing Sam's head.

"Whoa!" Sam yelped, ducking. "Leo, this is why we can't have nice things."

Leo doubled over in laughter. "Hey, at least we know it fails under extreme conditions. Scientific progress, my friends."

Graphene 2.0 and the Great Battery Experiment - Midway through the year, Leo's passion for graphene returned with a vengeance. He read about its potential applications in batteries and decided he'd create a prototype for a more efficient battery using a graphene composite.

Leo spent days working on the project, balancing the challenge of mixing graphene oxide with lithium compounds without blowing anything up. One evening, after many failed attempts that involved lots of smoke and very little success, he finally managed to create a small coin-sized battery. He hooked it up to an LED and flipped the switch.

The light glowed brightly for a moment, then dimmed. Leo frowned. "Well, it's progress."

Naveen wandered over, munching on an apple. "Hey, even Edison said he found 10,000 ways that didn't work. This is just number 9,999."

"Thanks, Naveen," Leo said, sarcastically patting him on the back.

They took turns testing the battery, laughing each time the light bulb flickered and died. By the end of the night, Leo added "Don't mix graphene and snacks" to their list of lab rules, after a small mishap where an apple peel fell into the mixture and they discovered the hard way that fruit and electrical components don't mix well.

The Great Polymorphic Prank and Its Unexpected Outcome - Materials science wasn't all serious work, of course. Leo's class decided to prank Professor Greenwell by placing polymorphic plastic on his chair. This special type of plastic, known for its ability to soften and reshape when heated, had been a topic of a recent lecture.

Leo added a twist by using thermochromic dye so the plastic would change colour when sat on. When Professor Greenwell walked into the classroom and sat down, his chair turned from dull gray to bright pink.

"Very funny," Professor Greenwell said, trying to hide his smile as the class burst into laughter. "Let this be a lesson in the application of thermoplastic polymers."

"Glad to see we passed!" Leo called out, earning a playful glare.

Project Night with Emily and the UV-C Cleaner - In one of the most ambitious projects of the year, Leo and Emily collaborated to create a UV-C light cleaner that could disinfect surfaces. It was based on real science: UV-C light can break down the DNA and RNA of microorganisms, effectively killing bacteria and viruses.

"I can't believe this isn't already standard in every public building," Emily said as she wired up the device.

Leo nodded. "I know. And if we get this right, it could be a big step forward in public health."

Their first prototype was a bit oversized and looked more like a robot with a flashlight taped to its head than a sleek, modern gadget. When they tested it, they both wore safety goggles and stood far back.

"It's like a sci-fi ray gun," Emily said, suppressing a giggle.

After a few tests and a minor electrical short circuit that caused the room lights to flicker ominously, they managed to make the cleaner work. The data they collected showed that the UV-C device successfully reduced bacterial growth on surfaces by nearly 90%.

"Imagine if we made this portable!" Leo said, excitement bubbling in his voice.

Emily grinned. "You mean like a lightsaber for germs? Let's save that idea for next year."

The Move to Third Year - Leo's second year came to an end with a final exam that involved designing a hypothetical space habitat using composite materials. Leo's project detailed the use of carbon nanotube-reinforced polymer walls, which could be both lightweight and extremely strong.

The exam, like the year, was tough but rewarding. Leo passed with flying colours and celebrated with his friends by biking down to a local café that served "Oxford's best" scones.

As they sat there, munching on scones and laughing over the year's stories-the failed polypropylene test, the pink chair, and the UV-C "ray gun"-Leo felt more confident than ever. Materials science was proving to be the perfect blend of challenging, inspiring, and endlessly fun.

With second year complete, Leo was ready for whatever adventures third year would bring, knowing that there were plenty more experiments to run, mistakes to make, and laughs to share.

Third-Year Breakthroughs, Internships, and Laboratory Mishaps

Leo's third year at Oxford began with the familiar excitement of crisp autumn mornings and bustling lecture halls. He was now a seasoned materials science student, and each semester only deepened his fascination with the subject. The field had everything: the allure of advanced technology, the creativity of problem-solving, and the occasional chance to blow things up (in the name of science, of course). Leo's final year at Oxford was a rollercoaster of excitement, stress, and late-night revelations that always seemed to come when he was half-asleep and hugging a mug of lukewarm tea. Third-year classes delved into more complex areas like nanocomposites, sustainable materials, and advanced metallurgy, but none of that intimidated Leo now. He had matured into a confident, quick-witted young scientist with the reputation of the "mad materials magician" among his peers.

The Intern Announcement - One rainy Wednesday afternoon, Professor Greenwell made an announcement that sparked a frenzy of whispers.

"Students, I have news. The Materials Research Lab in London is accepting interns for the upcoming term. This is a prestigious opportunity to work on real-world materials projects," he said, pausing dramatically. His eyes met Leo's. "And, I might add, it will look great on your CV."

Leo's pulse quickened. The Materials Research Lab was legendary. It had a reputation for being the birthplace of groundbreaking discoveries in materials science, from superhydrophobic coatings to advanced conductive polymers.

As soon as class ended, Leo made a beeline for the application forms. Sam raised an eyebrow as Leo shoved past. "Easy there, Mr. Eager. Are they handing out Nobel Prizes already?"

"Nope, just a chance to work at the lab that invented self-healing concrete," Leo replied, eyes sparkling. "I need this internship."

The Trials of Internship Application - Applying for the internship involved more paperwork than Leo had anticipated. Personal statements, letters of recommendation, a list of past projects-he even found himself writing a paragraph about his "biggest laboratory failure" and how it taught him resilience.

"Remember to add the 'sticky hydrogel incident'," Naveen called out, snickering from his desk.

Leo shook his head. "I'm not trying to scare them off, Naveen."

Finally, after weeks of nervous waiting, an email pinged into Leo's inbox: **"Congratulations! You have been accepted as an intern at the Materials Research Lab."**

Leo let out a whoop so loud that Emily, who was passing by his dorm room, poked her head in. "What's the fuss?"

"Internship, Emily. The Lab. London," Leo stammered, barely able to contain his excitement.

Emily's eyes widened. "That's huge! Just don't set anything on fire. Yet."

Internship Shenanigans: The Curious Case of the Nitinol Coil - Leo's first day at the Materials Research Lab was surreal. He walked past rooms filled with microscopes powerful enough to see atoms, machines that tested materials under extreme conditions, and scientists discussing experiments with phrases like "quantum weirdness" thrown around as casually as weather talk.

His supervisor, Dr. Patel, handed him his first assignment: studying **Nitinol**, a nickel-titanium alloy known for its shape-memory properties. Nitinol could "remember" its original shape and return to it when heated. Leo's job was to find new ways to apply this to medical devices.

"Think of it as metal with a brain," Dr. Patel said with a grin. "But don't ask it for opinions."

The challenge was to create a tiny Nitinol coil that would expand perfectly when exposed to body temperature. Leo spent days carefully bending, heating, and testing coils. On the third day, he made a breakthrough: the coil expanded smoothly when exposed to a heat lamp.

Feeling triumphant, Leo called over one of the senior scientists, Dr. Grayson, to demonstrate.

"Check this out," Leo said confidently as he flicked the heat lamp on. The coil began to move, but then, in an unexpected twist, it shot off the bench like a spring-loaded snake and hit the wall with a metallic *thunk*.

Dr. Grayson's eyebrows shot up. "Impressive. Not quite what we want in someone's bloodstream, but impressive."

Leo laughed nervously. "I'll... adjust the tension."

Classes, Laughs, and the Wonders of Aerogels - Balancing the internship with his university courses proved tricky. Leo's favourite class that semester focused on **aerogels**-incredibly lightweight materials made almost entirely of air. Aerogels have amazing insulating properties and are used in everything from space suits to window insulation.

One day in the lab, Leo and his friends decided to test the properties of a small aerogel sample by holding it over a Bunsen burner flame. Leo narrated dramatically, "And here, we observe the aerogel, undaunted by the heat. The miracle of modern science..."

As expected, the aerogel didn't catch fire, but it did start to crack apart. A tiny piece popped into the air and fell right into Leo's open water bottle.

Sam stared at the bottle. "So... do we drink it now for superpowers?"

Leo rolled his eyes and laughed. "Yeah, if you want 'super insulation' of your stomach."

Emily joined in. "What a groundbreaking discovery: flame-retardant water."

Breakthroughs and New Friends - The internship and classwork pushed Leo harder than ever. He met other interns who became friends and sources of inspiration, like Jess, a chemical engineering student who loved making puns.

"So, Leo, you're into Nitinol, eh? I guess you *shape* up nicely under pressure," she quipped.

"Only if I don't spring into a wall," Leo countered, earning a chorus of groans.

The Exhibition Project: A Humorous Exhibition - At the University's Science Exhibition, Leo's display featured his Nitinol work, complete with a demonstration. He added a twist: each coil was connected to a small speaker that played a note when it expanded, resulting in a metal symphony when the heat lamp activated.

Leo had perfected the Nitinol coil to expand predictably. Dr. Patel congratulated him and even mentioned including him in a future publication on shape-memory alloys.

As the coils played an off-key version of Beethoven's "Ode to Joy," the crowd erupted in laughter. Even Professor Greenwell, who was notorious for not cracking a smile, nodded in approval.

"Well, Leo, you've turned metallurgy into music," he said. "I'll give you points for creativity."

The Birth of Leo's Dissertation Idea - The idea for Leo's dissertation came during an evening brainstorming session with Sam, Emily, and Naveen. They were crammed into their dorm room, papers and snacks scattered everywhere. The topic? Combining sustainability with high-performance materials.

Emily tossed a crumpled piece of paper at Leo. "How about edible cutlery? You've already messed with edible aerogels."

Leo laughed. "Yeah, because nothing says 'appetizing' like dissolving spoon foam."

Naveen, looking up from a comic book, chimed in, "What if you made a material that's strong but biodegradable, something that could actually be used for construction but doesn't leave a mess behind?"

Leo's eyes lit up. "Naveen, you genius! A biodegradable, high-strength composite. What if we could use something natural, like cellulose nanofibers, mixed with a recycled polymer matrix?"

Sam's jaw dropped. "Wait, you're talking about a bio-composite material that could replace plastics or even some metals. That's… insane."

Leo grinned. "Exactly. It's crazy enough that it just might work."

Diving Into Research: The Cellulose Saga - The next weeks were a blur. Leo read about **cellulose nanofibers**, the structural component of plants that's lightweight yet stronger than steel when processed properly. The real trick would be finding a compatible recycled polymer to mix it with and ensuring the resulting composite was durable, biodegradable, and cost-effective.

He spent long nights in the lab, often surrounded by tangled fibres, beakers of dissolved plant material, and polymer goo. One memorable evening, he accidentally splattered some cellulose paste on his glasses. Emily, who came by to drop off notes, burst into laughter.

"Leo, is that a new anti-scratch coating or are you just trying out plant-based fashion?"

Leo wiped his glasses with a smirk. "Laugh all you want, but when these glasses can survive a drop from a plane, you'll want a pair too."

The Craziest Experiment: The Living Structure - As part of his dissertation, Leo proposed an outlandish idea: creating a small, load-bearing model bridge entirely out of his bio-composite. He dubbed it the "Eco-Bridge," and it was a mix of recycled PLA (polylactic acid) and cellulose nanofibers, infused with a touch of humour and sleepless nights.

Construction was more difficult than he anticipated. There was an infamous moment when a batch of the bio-composite hardened too quickly, sticking his fingers to a test mould. Dr. Patel, who happened to be visiting from the Materials Research Lab, walked in to find Leo desperately trying to free himself.

"Leo, are you attempting to *bond* with your project on a deeper level?" Dr. Patel quipped.

"Just testing the adhesive properties!" Leo grinned, finally wrenching his hand free and waving it triumphantly.

The Presentation Day: Laughter and Applause - Presentation day came faster than expected. Leo stood nervously beside his Eco-Bridge in a packed auditorium. His friends filled the front row, waving a homemade banner that read, "Leo: Bridging the Gap Between Science and Madness."

Leo cleared his throat. "Good afternoon, professors, fellow students, and anyone who's here just for the free snacks. Today, I present the Eco-Bridge: a biodegradable, high-strength composite structure that's eco-friendly and-if you're very desperate-technically edible. But please, don't eat it."

The audience chuckled, and Leo relaxed. He walked through the properties of the composite, detailing how cellulose nanofibres provided incredible tensile strength while the recycled polymer added flexibility and resilience. The bridge model, capable of bearing more weight than expected, was showcased using a tiny, comically overloaded toy truck rolling over it.

When Leo removed the weights and the bridge sprang back to shape, there was a moment of stunned silence, followed by thunderous applause.

Accolades and Appreciation - After the presentation, professors surrounded Leo, firing questions faster than he could answer. Dr. Patel clapped him on the shoulder. "Leo, this is exceptional. You've managed to combine sustainability and structural engineering in a way that could genuinely impact material science."

Professor Greenwell, who rarely gave compliments, nodded. "Top marks, Leo. This is the kind of innovative thinking we need more of."

Leo's project not only received the highest marks in the department but also sparked interest from outside the university. A local news article featured him as "The Student with an Eco-Friendly Edge," and companies reached out to learn more about the applications of his bio-composite.

A Night to Celebrate - The night after the presentations, Leo and his friends celebrated in the student lounge with pizza and mocktails. Sam lifted his glass. "To Leo: the only person who could turn plant fibres into fame."

"And to the future," Emily added, raising her glass. "Where bridges won't just connect people, but protect the planet too."

Leo beamed. He was exhausted, his fingers still smelled faintly of polymer, and he'd been awake for what felt like a week-but he was happy. This was what he'd always dreamed of: using science to make a difference, pushing boundaries while surrounded by laughter and friendship.

Moving to the Final Year - With third year completed, Leo felt proud but exhausted. Between his courses, internship, and impromptu dorm room experiments (including a brief, failed attempt to create "edible aerogels"), he knew he had grown immensely as a budding scientist and innovator.

As he and his friends celebrated with takeout pizza and one last prank (involving fake, sticky Nitinol coils that adhered to

Sam's chair), Leo couldn't help but feel excited for the year to come.

The path to the final year of his bachelor's degree lay ahead, filled with even more advanced topics, bigger challenges, and undoubtedly more moments of comic chaos. Leo was ready to dive in.

Third Year: Complete - Leo's final grades were posted a week later. Top of his class. His dissertation was officially archived as one of the most innovative projects in recent memory.

As he stood in the university courtyard, feeling the cool breeze and the weight of his accomplishment, Leo knew that materials science was more than a passion. It was his calling, one full of challenges, triumphs, and, above all, joy.

Third year was complete, and the future had never looked brighter.

Love, Laughter, and the Great Expedition

Leo's journey through college had been a whirlwind of discovery, late-night cramming sessions, triumphant experiments, and countless laughs shared with friends. He graduated with honours in materials science, and the accolades he received for his Eco-Bridge project paved the way for research opportunities and potential job offers. But Leo needed a break-a pause to reflect, re-energize, and explore the world beyond academic walls. So, with his suitcase stuffed with more lab goggles than T-shirts, Leo embarked on a journey of scientific exploration. His plan was simple: join various research expeditions and meet the scientists he had long admired, explore laboratories around the world, and maybe, just maybe, find the spark of inspiration that would guide the next chapter of his life.

Expedition #1: The Icelandic Geothermal Wonders - Leo's first stop was Iceland, famous for its geothermal energy research. As he arrived, the stark contrast between volcanic rocks and steaming vents fascinated him. He joined a group of scientists who were studying ways to harness this natural heat to develop eco-friendly energy solutions. They called themselves the "Steam Team," and Leo couldn't resist the pun-filled banter.

"Careful, Leo," said Dr. Asta, a veteran geologist, as she handed him a device to measure thermal conductivity. "The ground's hot, but not as hot as my volcanic jokes."

Leo chuckled. "I'll lava good pun any day, so fire away."

Their days were filled with sampling mineral deposits and experimenting with turbine models. One evening, Leo and the team tested a small-scale generator powered by geothermal steam. Just as the generator sputtered to life, emitting a triumphant whistle, Leo's phone buzzed with a notification: a reminder of his next expedition.

Expedition #2: The Antarctic Microbe Mission - Next, Leo flew to Antarctica, a continent that boasted breathtaking views of ice as far as the eye could see and temperatures that made his bones shiver. Here, scientists were exploring extremophiles-microbes that thrived in the most inhospitable conditions. The hope was to unlock the secrets of proteins that could withstand extreme cold, potentially leading to new materials with unprecedented durability.

Leo's task involved using a makeshift lab on a research ship to study the structure of these microbial proteins. One particularly comical day, Leo slipped on an ice patch, sending a petri dish flying. It slid across the deck, only to be caught by a fellow researcher whose face was hidden behind fogged-up goggles.

"Careful there, or you'll have runaway extremophiles," she said, lifting her goggles to reveal bright eyes and a warm, amused smile. The woman's name was **Mara**, a marine biologist whose passion for studying life in harsh conditions matched Leo's zeal for pushing materials science to its limits.

Leo felt an unexpected flutter in his chest. "Good catch. I was worried I'd have to name a new microbe after myself- 'Leoslipus Embarrassus.'"

Mara laughed, a musical sound that felt warmer than any parka. "That's one for the textbooks."

The Spark of Connection - Over the next weeks, Leo and Mara found themselves paired on numerous projects. They were the perfect blend of contrasts: Leo's focus on developing resilient materials meshed with Mara's insights into biological adaptations. One night, while analysing micrographs of ice-resistant microbial structures, they discovered that combining Leo's polymers with Mara's protein research could create a biocompatible, antifreeze material. The eureka moment came when their hands brushed reaching for the same printout, sparking a shared look that said, *Did we just make a breakthrough... and did something else just happen?*

"Is it me, or did the lab just get warmer?" Leo joked, trying to mask the rising colour in his cheeks.

Mara smirked. "Must be all this hot scientific progress."

They worked late into the night, crafting a prototype gel that, when applied to surfaces, could prevent ice formation without harming the environment. As the gel set, Mara sighed and leaned against the counter. "You know, Leo, I came here expecting to find microbes, not a partner-in-crime who turns petri-dish disasters into a comedy act."

Leo's heart raced. "Same here. Well, not exactly, I mean... microbes, yes. But I didn't expect to meet someone who could laugh at my science puns and keep up with them."

Mara's smile was as bright as the Antarctic sun. "I guess we're just *bonding* over shared interests."

The Humorous Expedition Conclusion - Word spread about their discovery, and soon the expedition team was eager to test the ice-proof gel on different surfaces, including the outside of the research ship. The test was successful, and the crew celebrated with a makeshift dance party in the mess hall, complete with ice cream sandwiches that Mara jokingly declared were "now immune to self-melting."

Leo and Mara laughed and danced, bumping elbows and tripping over each other in the cramped space. It was a far cry from a romantic dinner in a warm city, but it was perfect in its own chaotic, frozen way.

As the expedition wrapped up, Leo and Mara stood side by side, watching the last traces of the Antarctic sun dip below the horizon.

"So, what's next?" Mara asked, breaking the silence.

"Honestly?" Leo said, turning to her. "I have no idea. But I do know that I don't want the adventures to end."

"Good," Mara said, slipping her arm through his. "Because I hear there's a research station in Patagonia studying bioinspired materials. And I could use a partner who knows his way around cellulose nanofibers-and doesn't mind a bit of slipping on ice."

A New Chapter Begins - Leo knew his path had changed, expanding beyond the limits of laboratories and lecture halls. Science was still his passion, but now it came with laughter, shared glances over microscopes, and a future he hadn't planned but was eager to explore.

And as they boarded the plane to their next destination, surrounded by half-packed samples and laughter-filled conversation, Leo felt ready for the newest chapter of his life: one where innovation and heart went hand in hand.

Research, Romance, and Redefining Science

The journey to Patagonia was a whirlwind of cold drafts, shared headphones, and makeshift playlists that swung between classical scores for concentration and 90s pop hits for spontaneous dance breaks at the airport terminal. Leo and Mara had quickly become the duo everyone on the expedition recognized-the researchers who brought not just keen insight but laughter to every project.

Their destination was the **Andes Research Station**, a hub for biologists, engineers, and chemists working on bioinspired materials, particularly those mimicking the unique traits of regional wildlife. Leo's expertise in materials science aligned perfectly with Mara's knowledge of organic adaptations, and they were invited as collaborative leads to find solutions to age-old problems like ice buildup on infrastructure and materials that could withstand extreme temperatures.

The First Big Experiment - Leo and Mara spent their first few weeks setting up their lab space, which was an adventure in itself. One afternoon, Leo found himself tangled in wires and tubes while trying to rig a temperature-controlled bioreactor.

"Need some help, Professor Slip-and-Tangle?" Mara called from across the room, eyebrows raised in amusement.

Leo pretended to consider it. "Nah, I'm just trying to create a new knot-based polymer synthesis technique."

"Ah, the elusive *Cluelessus Maximus* method," Mara said, stifling a laugh as she handed him a screwdriver.

With their station finally assembled, they began experimenting with a unique polymer blend infused with proteins inspired by the scales of the **Andean mountain cat**, known for its incredible insulation properties. The goal was to develop a material that could maintain temperature consistency without excessive heating or cooling-a game-changer for both clothing and energy-efficient buildings.

A Day of Chaos and Chemistry - The day they ran their first big test, the lab was alive with whirring machines, clinking beakers, and the occasional muttered, "Why won't you polymerize, you stubborn molecule?" The experiment was going well until one of the bioreactors began to make a sputtering sound.

Mara, who was adjusting a microscope, glanced at Leo. "Is it supposed to sound like a bad beatboxing act?"

Leo's eyes widened. "Only if we're aiming for experimental hip-hop."

Before either could react, a geyser of blue gel shot from the bioreactor, splattering across the room and coating them both. There was a moment of stunned silence, broken only by the drip-drip of the gel from the ceiling.

Mara blinked, wiping gel from her hair. "Well, it's definitely *viscous*, and it sticks well. Future glue prototype?"

Leo, who had a streak of blue across his nose, burst into laughter. "We might have just invented the world's first thermally reactive party confetti."

The team members who came running at the sound of the commotion doubled over with laughter when they saw the state of their so-called lead researchers.

The Aftermath: Surprising Success - After cleaning up, Leo and Mara found that the spilled polymer mixture had bonded into an unexpected, resilient sheet on the lab table. They ran tests and discovered that the accidental bonding had resulted in a substance that could retain heat and resist moisture better than their original designs. It was a breakthrough that led to a prototype for a self-insulating, waterproof material.

Their supervisors were delighted, and the pair's reputation for quirky success stories grew. Colleagues from other stations would stop by just to see what "Team Accidental Brilliance," as they'd been nicknamed, was up to.

The Spark Grows - As weeks turned into months, Leo and Mara's partnership deepened beyond the confines of the lab. They spent evenings under the clear, starry skies of Patagonia, sitting on rocky ledges with warm cups of mate in hand. One particularly chilly night, Mara pointed to the sky.

"See that constellation?" she said. "That's Crux, the Southern Cross. My grandfather used to say it points toward new journeys."

Leo smiled, the cold air forming a halo of mist around them. "Well, he wasn't wrong. It's kind of poetic, don't you think? We came all this way to study materials and ended up discovering each other."

Mara turned to him, eyes sparkling even in the dim light. "You're getting better at these lines, Leo."

"I've had a great teacher," he quipped, leaning in just as the wind picked up, sending a gust that made them both laugh as they huddled closer.

Their First Joint Presentation - The final experiment results led to an invitation to present their findings at a prestigious conference in Santiago. Leo and Mara prepared an engaging, humour-infused presentation that showcased their revolutionary bioinspired insulation material. The title slide read, "From Cat Scales to Clumsy Science: A Tale of Unexpected Inventions."

The room was filled with scientists, industrial leaders, and students. Leo started, "Welcome, everyone. Before we begin, I'd like to assure you-our lab's safety measures are as strong as our new polymer, even if our anecdotes say otherwise."

Mara followed up, "And if you hear about a blue confetti incident, please remember that all discoveries start with a splash...literally."

Their blend of humour and cutting-edge science was met with laughter and applause. The audience hung on to their every word, inspired by how science and serendipity could create innovations that had real-world impact.

A Future Full of Science and Love - After the conference, Leo and Mara returned to the research station with more than just newfound recognition-they had a deeper sense of partnership, both professional and personal. They joked that their life was a series of controlled reactions with unpredictable outcomes, and they wouldn't have it any other way.

As they sat one evening, sketching out ideas for their next big project (a new adhesive inspired by spider silk), Leo turned to Mara. "What do you think comes next for Team Accidental Brilliance?"

Mara smiled, leaning into him. "More adventures, more accidental successes, and definitely more stories worth telling."

Leo couldn't agree more. Science was an endless journey, and with Mara by his side, every discovery was an adventure waiting to happen.

From Conference Applause to Life's Big Pause

Leo and Mara's return to the Andes Research Station after their successful presentation in Santiago felt like a victory lap, with the rest of the team greeting them like scientific rockstars. Leo, never one to shy away from self-deprecating humour, made a show of pretending to sign autographs on leftover lab gloves, eliciting laughter and groans in equal measure.

"Leo, stop inflating your ego, or we'll need an anti-bloat polymer," teased Dr. Asta, who had become a close friend since their Iceland days.

Mara laughed as she passed Leo a clipboard. "Careful, Dr. Slip-and-Tangle. The next experiment might involve ego shrinkage."

Leo raised an eyebrow. "An inverse polymer! We'd make millions."

The Invention of Spider Silk Adhesive - The following weeks saw Leo and Mara's lab filled with diagrams, sticky notes, and the faint smell of burnt polymer (which Leo insisted was part of the creative process). Their new project aimed to mimic the natural adhesive properties of spider silk-a material known for being lightweight, strong, and capable of sticking to almost anything without being too difficult to remove. Leo called it their "Spidey Sense of Materials."

One memorable experiment had Leo trying to spin the synthetic silk around a small model of a bridge they'd built for testing. The thread stuck perfectly at first, but then snapped back with an unexpected force, sending a tiny scale-model car flying across the lab. It landed in Mara's coffee cup with a resounding splash.

Mara sighed, holding up the dripping toy. "I don't remember 'latte with car' being on the menu."

Leo tried to look serious. "Prototype transport adhesive: a work in progress."

The tests and failures led them to experiment with a combination of carbon nanotubes and synthetic silk proteins, creating an adhesive that maintained strength but could be easily dissolved with a mild solution. This discovery had potential applications in fields from medicine (temporary stitches) to construction (lightweight, strong scaffolding).

Life Outside the Lab: Humour in the Ordinary - Despite their intense research schedule, Leo and Mara made sure to carve out time for activities that didn't involve lab coats or safety goggles. They explored the Patagonian terrain, hiked the rugged paths, and shared quiet nights by campfires where Mara recounted stories of marine biologists who accidentally discovered new species while looking for their lost glasses.

One day, Leo decided they needed a break with more excitement. "Let's go glacier kayaking," he suggested.

Mara looked sceptical. "Is this another one of your ideas that ends with a splash?"

Leo winked. "Guaranteed. But this time, we're wearing wetsuits."

Their trip involved navigating icy waters that glistened with sunlight, creating a surreal landscape. The serenity was briefly interrupted when Leo's paddle caught on a chunk of floating ice, sending him teetering sideways.

"Don't do it!" Mara shouted, her laugh barely contained.

Too late. Leo tipped over with a splash, only to pop up moments later, drenched and grinning. "Note to self: next research project-buoyant, self-righting paddles."

Recognition Beyond the Research Community - Word of Leo and Mara's adhesive invention spread beyond the scientific community. Articles with headlines like *"Adhesives with a Twist-The Duo That Sticks Together"* appeared in science magazines and blogs. The university even received queries from companies interested in licensing the technology. While both Leo and Mara were flattered by the attention, they were adamant about ensuring the material would first benefit conservation projects, like lightweight constructions in disaster-stricken areas.

Professor Greenwell, their former mentor from Oxford, reached out. His email read: "I always knew you'd turn the unexpected into the groundbreaking. Never change, Leo. Also, avoid glacier kayaking-I speak from experience."

The Proposal (And Another Splash) - A few months later, on a warm summer evening, Leo stood on the shores of Lake Pehoé, the vibrant turquoise water shimmering in the light. The research station staff had gathered for a barbecue, and the air buzzed with laughter and the smell of grilled vegetables and local fish. Leo had planned something special. Mara was by his side, unaware of the small box in Leo's pocket.

As the sun dipped below the horizon, Leo nudged her and said, "Hey, Mara. Remember when we started with runaway petri dishes and polymer mishaps?"

Mara laughed, eyes sparkling with memories. "How could I forget? It's practically our origin story."

Leo took a deep breath, then dropped to one knee. The sudden movement made a nearby folding chair collapse with a squeak, drawing attention from their friends.

"Are you-?!" Mara's voice caught as she realized what was happening.

Leo, eyes locked on hers, said, "Every discovery has been better with you by my side, whether it was planned or accidental. Mara, will you be my partner, not just in the lab, but in life?"

A collective "aww" rose from the crowd, quickly followed by shouts of, "Say yes!" and "About time, Leo!"

Mara's eyes brimmed with tears as she nodded. "Of course, you absolute goof."

They kissed as the sun disappeared, and just as they did, a splash came from the lake behind them. One of their more daring friends had thrown a rock into the water, setting off a series of ripples and laughter.

Leo turned and shouted, "That's not how you make a splash in science, but I'll take it!"

The Future Unfolds - With their engagement, Leo and Mara's life only became more intertwined. They took on speaking engagements, blending humour and scientific insight, captivating audiences from university students to CEOs. Their lectures were full of stories about "exploding bioreactors" and "antifreeze gel battles," each tale punctuated by an unspoken promise that science could be as full of fun as it was of breakthroughs.

As they prepared for their next major project-an exploration into eco-friendly building materials that could self-repair using natural proteins-Leo leaned against their lab table, watching Mara measure a solution.

"Ready for another round of controlled chaos?" he asked, eyes twinkling.

She looked up, smirking. "With you? Always."

And with that, they dove into a future as bright and unpredictable as the discoveries they were about to make, bound by science, love, and laughter that echoed through the halls of every lab they stepped into.

Settling into the Spotlight

Leo and Mara found themselves in a new chapter, basking in the warm glow of their recent success and the thrill of their engagement. Their groundbreaking adhesive continued to attract attention, not only from commercial entities but from environmental conservationists who saw its potential to revolutionize lightweight, eco-friendly construction in disaster zones. Invitations to conferences, research collaborations, and guest lectures piled up like a precarious tower of lab reports on Leo's desk.

One day, Mara glanced at the heap and quipped, "This pile might collapse under its own gravitational force. Shall we classify it as a new planet?"

Leo smirked. "If it does, I'll name it 'Paperopolis' and apply for funding to study its properties."

Despite their growing fame, they managed to stay true to their quirky essence. Leo's fondness for science puns became legendary in their talks. At one presentation on polymer resilience, he joked, "The key to our discovery? We didn't just 'stick' with conventional methods-we 'adhered' to the wackiest ideas we could."

New Horizons: The Self-Healing Material - Eager to push the boundaries, the duo embarked on their most ambitious project yet: developing a self-healing material capable of repairing structural damage. The inspiration came from *biomimicry*, particularly observing the remarkable self-repairing qualities of skin and plant tissue.

"Think about it," Leo said during a brainstorming session. "Nature figured out how to heal millions of years ago. We just need to take notes."

They started by examining the *mechanism of self-healing hydrogels*, materials that could bond at a molecular level to repair minor damage. The basis of their work lay in incorporating polymers with reversible covalent bonds and microcapsules containing a healing agent. When the material cracked, the capsules would rupture, releasing the agent to mend the break.

In one experiment, Leo tried to demonstrate the concept on a scale model of a bridge built from their material. Everything seemed fine until Mara leaned in to inspect a crack, and the microcapsules burst prematurely, releasing a bright purple substance that squirted across her face.

Leo stifled a laugh. "Oh no, it's the new Mara-makeup line: *Grape Regenerate*."

Mara, purple-streaked but grinning, replied, "If this works, I'll name it 'Mara-velous Mend'."

Their hard work paid off. Within months, they showcased their self-repairing material at a global conference in Zurich. Leo's demo featured a dramatic moment where he snapped a thin piece of their polymer sheet, only for the audience to watch in awe as it fused back together over the course of a few minutes.

The room erupted in applause. One of the panelists leaned over to another and whispered, "They're not just making materials; they're redefining them."

Life Among Labs and Laughter - While their professional life soared, their personal life kept them grounded. They still managed to find humour in the mundane-like the time Leo thought he'd created a new insulating foam but ended up with a substance that expanded so rapidly it filled their lab in a matter of minutes.

Mara, peeking through a window, shouted, "Looks like the lab has a new marshmallow-themed interior!"

Leo, waist-deep in expanding foam, shouted back, "At least it's thermally efficient!"

Their friends from the station, including Dr. Asta and their old Oxford mentor Professor Greenwell (who made surprise visits), became regulars at their impromptu barbecues, where discussions of the latest scientific findings were punctuated by playful debates about who would survive in a *post-apocalyptic scenario driven by sentient polymers*.

A Love Story Written in Science - Leo and Mara's relationship blossomed through their work and shared laughter, culminating in a wedding that was both unconventional and scientifically themed. Held at the research station's courtyard, the ceremony featured test tubes filled with wildflowers and a cake decorated with hexagonal patterns symbolizing molecular bonds.

During his vows, Leo couldn't resist a science reference: "Mara, in the lab of life, you're the catalyst that makes everything better. I promise to always be your reaction partner, even if the outcomes are unpredictable."

Mara's eyes glistened as she replied, "And I vow to always share my findings, my joy, and the occasional splash of blue gel with you."

Their vows were met with laughter and cheers, and as they sealed their promise with a kiss, Dr. Asta, always the joker, tossed a handful of synthetic glitter with the words, "May your bond never denature!"

A New Adventure: Fieldwork in the Amazon - Fresh from their wedding, Leo and Mara were invited to join an interdisciplinary team heading to the Amazon rainforest to study the *structural properties of natural composites*, such as those found in the strong yet flexible fibres of local vines. These fibres had the potential to inspire sustainable building materials with low environmental impact.

Their arrival in the Amazon was met with awe and an onslaught of biting insects. Leo, swatting a mosquito, joked, "I hope they appreciate my contributions to their protein intake."

Mara laughed. "If this keeps up, we'll need to patent 'Mosquito Armor 2.0.'"

The team spent days hiking, gathering samples, and analysing the biome's secrets. Leo's sense of wonder only grew as they encountered bioluminescent fungi and observed the *structural composition of spider webs*, known for being stronger than steel by weight. It was a treasure trove of ideas for future projects.

One evening, as they sat by their tents under a sky streaked with stars and the hum of the rainforest around them, Leo turned to Mara. "We started with polymers and purple splashes. Who knew we'd end up here, chasing science through the wild?"

Mara leaned in, eyes full of affection. "I wouldn't have it any other way."

Recognition and the Call for New Challenges - The Amazon expedition yielded groundbreaking research, leading to new sustainable materials that mimicked the strength and flexibility of natural fibres. Articles about their work appeared in major journals, complete with a cheeky illustration Leo submitted: a comic of a mosquito with a speech bubble saying, "Thanks for the buffet, Leo."

Universities around the world invited them to give talks and join collaborative projects. Yet amidst the accolades, Leo and Mara knew that their passion lay not just in their achievements but in the journey itself-full of trial, error, and the occasional lab mishap that ended in laughter.

Their story continued as they embarked on more fieldwork, mixing humour with their unending curiosity. Leo once summarized their approach in an interview, saying, "If we can't have fun in the process, are we really scientists? Discoveries are great, but joy is the best byproduct."

The interviewer noted, "It seems like the world could use more of your kind of science."

Leo's response was classic: "Just remember, never underestimate the power of happy accidents."

The Global Stage and Humbling Returns

With their Amazon research making headlines, Leo and Mara became household names in scientific circles. Invitations to conferences flooded their inbox, from Dubai's Future of Materials Symposium to Tokyo's EcoTech Summit. Despite the glamorous offers, they chose their engagements carefully, ensuring they had time to return to the essence of their work-hands-on research and teaching. One of their most impactful trips was to the University of Tokyo, where they were invited to speak about the *biomimicry of Amazonian fibres*. Leo, never missing an opportunity for a quirky icebreaker, started the talk by holding up a tangled vine and saying, "This little guy once saved me from falling into a very unkind mud pit. Nature, folks-better than any gym membership."

The audience chuckled as Mara stepped in with, "And that's how we discovered our next project. Leo's missteps are research in disguise."

The discussion then turned to the structural analysis of the vine's fibres, which revealed a natural cross-linking process that inspired them to develop a synthetic version with applications in biodegradable construction materials. Leo's humour and Mara's depth of explanation won the crowd over, leaving an impression beyond scientific facts.[1]

---

[1] Many natural fibres, like those found in the Amazon, possess hierarchical structures that distribute load efficiently. Materials scientists use this as inspiration for creating synthetic fibres with similar properties.

A Prank-Worthy Lab Experiment - Back in their own lab, Leo decided to lighten the mood by pranking their research team. Using non-Newtonian fluids-a substance that acts like a liquid when poured and a solid when under pressure-he set up a tray filled with cornstarch and water at the entrance. When Dr. Asta walked in, she stepped on it, expecting a solid floor but finding herself sinking slightly before the material held firm.

"Leo!" she yelled, balancing precariously. "What is this-quick-mud?"

Leo, holding back laughter, called out, "A little demonstration of shear-thickening behaviour! Your reactions are more dramatic than the stress-strain curve."

Mara, who had watched it unfold from her desk, added, "Leo, you know payback in a science lab involves controlled chaos, right?"

The incident became a legend in the lab, leading to a semi-serious "Pranks and Practical Physics" board where staff logged funny experiments gone awry. [2]

From Fun to Fortune: A New Material - Leo's humorous nature belied his relentless curiosity. Late one night, tinkering with their self-healing polymer and the properties they discovered in the Amazonian fibres, Leo accidentally created a prototype that exhibited *thermo-responsive behaviour*. This meant it expanded or contracted based on temperature changes-a feature with implications for wearable tech and thermal insulation.

---

[2] Non-Newtonian fluids like cornstarch and water display shear-thickening behaviour, meaning they become more solid under pressure but remain liquid otherwise. This property is explored for applications in protective gear.

He rushed to Mara with the prototype. "Mara, meet our new friend: *ThermoBuddy*," he said, holding up a swatch of what looked like a thin, flexible film.

Mara, intrigued, took the material and ran a quick test. "Leo, this could revolutionize adaptive clothing. Imagine a jacket that adjusts its insulation depending on the weather."

Leo's eyes lit up. "Or socks that stop you from freezing during those outdoor expeditions. No more cold feet-scientifically speaking."

The ThermoBuddy prototype gained traction, and soon they were in talks with tech companies for applications in clothing lines, solar panel insulation, and even lightweight habitats for harsh environments.[3]

The Unexpected Visit - One morning, while testing their ThermoBuddy material for durability, a knock at their lab door startled them. It was Professor Greenwell, now retired but still as lively as ever. He peered through his round glasses and exclaimed, "I've come to see what my two star pupils are up to. I heard rumours about socks with attitudes."

Mara welcomed him with a hug. "Professor, we've missed your metaphors! How's retirement?"

He grinned. "Too peaceful, so I thought I'd disrupt yours. Show me the magic."

Leo demonstrated their latest discovery, and Greenwell's eyes sparkled. "This is brilliant! But, have you considered the reversible phase transition property? It could stabilize under drastic temperature shifts."

---

[3] Thermo-responsive materials expand or contract when exposed to temperature changes, due to the polymer chains' reaction to heat or cold. This concept has potential in various fields, including energy-efficient building materials and smart textiles.

Leo slapped his forehead. "Why didn't we think of that? Mara, note this down. Professor, you've just upgraded our ThermoBuddy to ThermoBuddy 2.0."

The visit turned into a brainstorm session, and for a brief moment, it felt like their days at Oxford all over again. Leo couldn't resist joking, "Next time, we're naming something after you. Maybe a self-patching umbrella."

The Proposal with a Twist - As the ThermoBuddy material gained more recognition, Leo and Mara's popularity soared. They were asked to deliver a joint keynote speech at a prestigious global science expo in Geneva. The speech, full of anecdotes, science demonstrations, and witty banter, was a resounding success.

Afterward, in the garden behind the expo centre, Leo surprised Mara with a small box. Inside wasn't a new gadget, but a bracelet made from a piece of their original self-healing polymer.

"Mara, in this bracelet is our story-the stumbles, the breakthroughs, and the purple splashes. Will you wear it as a reminder that whatever comes next, we're ready for it?"

Tears welled up in her eyes as she put it on. "Leo, this might be your best invention yet."[4]

Global Collaboration and New Discoveries - Leo and Mara's research attracted the attention of international labs eager for collaboration. They embarked on a joint project with a team from Germany specializing in *piezoelectric materials*, substances that generate electricity when pressure is applied. The goal? To integrate piezoelectric properties into their

---

[4] Self-healing materials are polymers that can repair damage autonomously through chemical reactions or physical properties triggered by changes in temperature, pressure, or light.

ThermoBuddy material, allowing clothing that could generate small amounts of power from everyday movement.

The experiments were rife with funny moments. Leo wore a test jacket and danced around the lab to measure energy output, proclaiming, "Behold, the first dance-powered phone charger!"

Mara recorded the whole thing. "Let's make sure to keep this video. It's blackmail for when you get too serious."

Their final prototype-a thin, flexible material with integrated energy-harvesting capabilities-was a breakthrough. With potential applications in self-sustaining wearable tech and disaster relief clothing, it marked a new chapter in their work.[5]

Adventures, Challenges, and Everlasting Curiosity - Leo and Mara's adventures continued, from participating in international expeditions to teaching workshops on sustainable materials. Each new venture brought its share of missteps and marvels. They even hosted an online series called "Lab Laughs and Lessons," showcasing the lighter side of scientific research.

One episode featured a segment where Mara accidentally mixed the wrong solution, resulting in harmless, bubbly foam. Leo leaned into the camera and said, "And this, folks, is how we discovered 'Foam-o-rama,' the latest in stress-relief toys."

---

[5] Piezoelectric materials produce electric charge under mechanical stress. They're used in everything from medical sensors to energy-harvesting devices.

Lab of Infinite Surprises

Leo and Mara's new lab had become a haven for innovation and hilarity. Their collaborative work with German scientists on piezoelectric fabrics was gaining traction, prompting a visit from Dr. Heike, a renowned materials scientist known for his stoic demeanour and precise manner. Leo, always aiming to break the ice, wore one of their energy-harvesting jackets to greet him.

"Welcome, Dr. Heike! This jacket can power a small flashlight when I do this," Leo said, breaking into a goofy dance that had Mara burying her face in her hands.

Dr. Heike's lips twitched-almost a smile. "Interesting. Does it power your logic circuits as well?"

The whole room burst into laughter, even Dr. Heike, whose humour was usually as rare as stable antimatter.[6]

Expedition to the Arctic - Their next project took them north, joining a research group in the Arctic to test the performance of their ThermoBuddy material in extreme cold. The cold-climate tests involved strapping sensors to their gear, which would measure heat retention and material resilience.

Leo found himself in an oversized parka, surrounded by ice floes and playful seals. "I never thought I'd see the day when

---

[6] Piezoelectric fabrics convert mechanical stress into electrical energy. Such fabrics can be used to create self-powering devices like heart rate monitors embedded in clothing.

I'm the warmest thing in sub-zero temperatures," he joked, steam puffing as he spoke.

Mara, brushing snow off her goggles, chuckled. "And here I thought the most extreme thing about our work was navigating conference buffets."[7]

One evening, as the sun dipped low over the horizon and painted the ice with hues of orange and pink, Leo leaned back and whispered, "Even with the freezing toes and the frost on my beard, this is worth it."

Mara nudged him. "Plus, your dance moves double as warmth generators."

Leo raised an eyebrow. "Mara, we might have to market that as 'Scientific Swing.'"

The Competition of Quirks - Back in their lab, they decided to host a competition called "The Quirk-Off," where researchers showcased their most unconventional inventions. Among the entries were a harmonica that powered LED lights through breath-generated energy and a "dancing plant" that swayed thanks to micro-electrical pulses mimicking natural light movement.

Leo's entry was a robot shaped like a sloth that used artificial intelligence to find the warmest spot in any room. He dubbed it "HeatLazy," and it slowly rolled across the floor, only to nestle by the window where sunlight streamed in.

---

[7] Extreme cold tests on materials check for thermal conductivity and insulation properties. Insulative fabrics slow down heat transfer, critical for survival gear in polar environments.

Mara couldn't stop laughing. "This robot is basically you during winter mornings."[8]

The competition ended with awards that were equally whimsical-Leo won "Most Relatable Robot," while another team took "Most Musical Device" for their light-up harmonica. It was a day full of cheers, applause, and the occasional foam explosion from a mis calibrated presentation.

A Conference with a Twist - The duo's reputation led them to keynote at an international materials conference in Barcelona, where Leo decided to present in his unique fashion. He wheeled out a cart covered by a tarp and revealed a model of a suspension bridge made entirely from their self-healing, piezoelectric polymer.

"Now, imagine your bridge repairs itself after damage and powers streetlights from the vibrations of passing traffic," Leo said, tapping the model. The audience leaned in, fascinated.

Mara stepped in, a playful gleam in her eye. "And no, we won't claim it could withstand the dance moves Leo showcased last year."

The audience erupted into laughter, and even the sternest professors exchanged amused looks. The presentation ended with a standing ovation and a swarm of interested researchers eager to learn more.[9]

The Day of Surprises - One of their most unexpected challenges came when a leak in the lab's water system caused

---

[8] Robotics inspired by animals, known as *biomimetic robots*, incorporate movement and behavioural patterns from living creatures. This approach leads to more adaptive, versatile machines.

[9] The concept of self-healing infrastructure, especially bridges and roads, uses embedded microcapsules or shape-memory polymers that respond to stress or damage by releasing a healing agent.

a small flood. Leo, knee-deep in water and holding a clipboard over his head, shouted, "Well, now we can test ThermoBuddy's waterproof properties in real-time!"

Mara, wading through the water with an amused smirk, said, "Leo, I swear, if we accidentally invent something from this, it'll be called 'Flood-formance Fabric.'"

The mishap resulted in a surprising discovery. The material's water-resistant properties were even better than expected, which opened up a new line of research into moisture-wicking and waterproof clothing for extreme environments.[10]

A Love of Science and Life - Even with their busy schedules, Leo and Mara made time for each other and their friends. They held themed dinners where the menu items were named after famous scientific principles, like "Newton's Apple Pie" and "Curie's Quiche."

One evening, during a toast, Dr. Asta raised her glass and said, "To Leo and Mara, who prove that science isn't just about results but about enjoying every ridiculous step along the way."

Leo grinned. "Especially if that step lands you ankle-deep in a non-Newtonian fluid."

Mara smiled, eyes reflecting both joy and gratitude. "Here's to a lifetime of laughter, discoveries, and unplanned foam parties."

---

[10] Waterproof fabrics often incorporate hydrophobic coatings or special polymers to repel water while allowing breathability. Their development is key in sports and outdoor gear.

Renaissance Fair of Science

Leo and Mara were invited to a unique event called the "Renaissance Fair of Science," a celebration where scientists showcased inventions inspired by historical ingenuity. Leo had worked tirelessly to build a trebuchet that launched biodegradable balloons filled with safe, coloured foam—a nod to both medieval engineering and modern eco-friendly materials.

Mara, dressed as a 16th-century alchemist, stood next to a demonstration of a water purification device inspired by the Archimedes' screw but made with modern self-cleaning polymers. The fair was a mix of laughter, awe, and plenty of foam.

As Leo launched the trebuchet, the balloons burst in midair, creating a colourful display. "It's both a spectacle and a way to test polymer durability under stress!" he declared.

One balloon veered off course and hit Dr. Heike, who had attended as a guest judge. Silence fell before Dr. Heike wiped the foam from his glasses and said, "Leo, I see your experiments are never just confined to the lab."

The crowd roared with laughter.[11]

The Unexpected Team Building Exercise - Leo and Mara's research lab decided to hold a team-building weekend involving puzzles and physical challenges to foster

---

[11] A trebuchet operates using a counterweight that, when dropped, propels the arm and launches the payload. The mechanics of leverage and energy transfer are a testament to early engineering that modern engineers still study for principles of physics.

collaboration and reduce stress. The twist? All activities were based on scientific principles. For one challenge, teams had to build makeshift catapults using only rubber bands, popsicle sticks, and marshmallows.

Leo's team named their catapult "Marsh-Marshal" and tested it with much enthusiasm. The first attempt resulted in a marshmallow launching backward and landing on Dr. Asta's coffee cup. "Leo, it seems your kinetic calculations were slightly... inverted."

Mara, leading the rival team, smirked. "I think I'll call our creation 'Precision Pop.'" Her team's catapult launched perfectly, and the marshmallow landed directly in the target cup.

Leo clapped. "You win this round, Mara, but only because I was distracted by thermodynamics. Marshmallows absorb heat, you know."[12]

Discovering Graphene's Quirks - In their ongoing research, Leo and Mara delved into the properties of *graphene*, a single layer of carbon atoms arranged in a two-dimensional lattice. Leo was fascinated by its strength and electrical conductivity, imagining the applications in everything from flexible screens to energy-efficient batteries.

"Did you know, Mara," Leo said one day, peering at a microscopic sheet of graphene, "a single sheet is 200 times stronger than steel and can conduct electricity better than copper?"

---

[12] Catapults work on the principle of stored potential energy in an elastic band, which converts into kinetic energy when released. These basic principles are taught in mechanics and are applicable to modern-day energy storage devices.

Mara raised an eyebrow. "Are you suggesting we replace our lab chairs with graphene sheets?"

Leo grinned. "Only if we can make them self-heating and unbreakable."

One day, while demonstrating its strength, Leo accidentally cracked a pen that had a piece of graphene embedded in its casing. Ink splattered everywhere, marking Leo's face and lab coat. Mara laughed so hard, she nearly dropped her own research notes.

"Graphene: strong enough to defy pens but not Leo's coordination," she said, chuckling.[13]

The Puzzle of Smart Textiles - Leo and Mara's latest project involved integrating graphene into *smart textiles* to create clothing that could monitor vital signs, charge devices, or even change colour with environmental conditions. The project led to humorous moments when Leo wore a prototype shirt that glowed when he got excited, like when he discovered leftover pizza in the breakroom.

Dr. Asta noted dryly, "Leo, your enthusiasm is literally visible now. Useful for science talks and surprise parties."

The most intense test came during an impromptu lab fire drill. The shirt, designed to respond to changes in temperature, turned a bright, alarming red. "Guess it works," Leo said, waving to the amused team while outside.[14]

---

[13] Graphene is indeed an exceptional material with a tensile strength far surpassing steel, electrical conductivity superior to copper, and potential uses in many technological fields, including electronics and materials science.

[14] Smart textiles use conductive materials like graphene or embedded sensors to interact with external stimuli, making them capable of monitoring health, generating energy, or responding visually to temperature and stress changes.

Mara's Ingenious Experiment

Mara took charge of a project focusing on *shape-memory alloys*—metals that return to a pre-defined shape when exposed to heat. She created a small sculpture that twisted and moved when a hairdryer was pointed at it. Leo nicknamed it "Heat-Dancer."

During a demo for visiting scientists, Leo switched on the hairdryer, but the alloy responded by forming a shape that looked suspiciously like a cat.

Mara whispered, "Leo, did you program this thing for a feline response?"

He shrugged. "Just testing its artistic potential."

Their Largest Breakthrough Yet - Leo and Mara's collaborative work led to an unexpected breakthrough. By combining the piezoelectric properties of their earlier projects with graphene's conductivity, they developed a *flexible power-harvesting mat*. The mat generated electricity from footsteps, opening possibilities for sustainable energy in public spaces.

Testing it at a busy shopping centre, Leo shouted as he saw the voltage meter spike with each passing person. "This mat turns 'walking power' into literal power!"

A passerby, intrigued, asked, "Does dancing on it double the energy?"

Leo, without missing a beat, said, "Only if it's scientifically choreographed."[15]

---

[15] Piezoelectric materials generate electrical energy when mechanical stress is applied. Combining this with conductive graphene increases the efficiency and durability of energy-harvesting devices, ideal for urban and wearable tech applications.

The Proposal for the Future - In a heartfelt and unexpected twist, Leo crafted a ring for Mara out of their piezoelectric polymer and graphene mix. "Mara," he said during a quiet evening in the lab, "this is powered by every adventure we've had. Will you be the constant in my variable life?"

Mara, eyes twinkling, nodded. "Only if we patent this as the 'Evercharge Engagement Ring'."

Leo laughed. "Deal."[16]

---

[16] The integration of energy-harvesting materials with everyday objects, such as rings or other wearables, showcases the move toward seamlessly integrating renewable energy technology into daily life.

Great Miscalculation

Leo and Mara embarked on a new, ambitious project focusing on *transparent solar panels* made from a unique combination of perovskite and graphene. The idea was to make windows that could double as power sources, harnessing solar energy without sacrificing visibility.

"Imagine every skyscraper generating its own energy," Leo said, adjusting his safety goggles as he prepared to test a sample.

As they unveiled their prototype at a demonstration, Leo leaned a bit too hard against the table, causing it to wobble. The panel toppled and landed on Dr. Heike's clipboard, leaving a perfect, solar-absorbing imprint.

Dr. Heike glanced at his now-power-generating notes and smirked. "Leo, you seem to have invented *photovoltaic paperwork*."[17]

The Test of Flexibility - One day, Mara decided to experiment with *flexible batteries* that could be incorporated into clothing. Leo, always eager to take things a step further, suggested embedding the batteries in socks. "Think of the practicality, Mara! Your feet can charge your phone while you walk."

Mara raised an eyebrow. "Leo, I'll support this only if you agree to be the first tester."

---

[17] Transparent solar panels leverage materials like perovskite, which can be layered in ultra-thin films. By altering the way light interacts with the surface, they create panels that are nearly invisible yet still harvest energy.

During the first trial, Leo walked around the lab with one sock glowing faintly blue, but the other remained dark. He tripped over a stool as he tried to figure out why only one sock was working.

"Looks like the right foot is more 'positive' about this idea than the left," Mara joked, helping him up.[18]

The Lab Ghost - As the research became more intense, they found themselves working late nights in the lab. It was during one such night that they began to hear strange creaks and hums.

"Must be the HVAC system," Leo said, but his voice wavered.

Mara, never one to let a joke slide, whispered, "Or maybe it's the *ghost of failed experiments.*"

The next creak made Leo jump, knocking over a model of their piezoelectric mat, which, in the dark, emitted a soft, eerie glow.

"I'm starting to think we accidentally built a haunted mat," Leo said, wide-eyed.

They later discovered that the sounds were from a loose screw in the air duct and had a good laugh about their overactive imaginations.[19]

An Unexpected Invitation - Word of their work began to spread beyond the UK, and Leo and Mara received an invitation to present at an international *Materials Science Symposium* in

---

[18] Flexible batteries are built using components like graphene and polymer electrolytes, which allow them to bend and flex without losing conductivity or energy capacity. They're crucial for the future of wearable technology.

[19] Piezoelectric materials produce electrical charge when subjected to mechanical stress. When combined with LED lights, even the slightest pressure can make them glow, which can create spooky effects in dim environments.

Japan. It was an opportunity to showcase their transparent solar panels and flexible battery tech.

Their trip was filled with new experiences—Mara marvelled at the efficiency of Japanese high-speed trains, while Leo tried to order lunch using an electronic menu but somehow managed to end up with five bowls of rice.

"Looks like you're fuelling up for the whole week," Mara teased, pointing at the mound of rice.

Leo shrugged. "More energy for the brain, right?"

The presentation was a resounding success. Leo demonstrated the solar-absorbing windows by positioning them in front of a projector, and Mara explained the science behind their battery-embedded fabrics. They were met with applause and a few questions about when their inventions would be commercially available.[20]

An Unlikely Partnership - Back home, they received an email from a global sportswear company interested in collaborating to create self-heating and energy-generating winter wear. Leo and Mara were thrilled at the prospect and immediately started brainstorming.

"I'm thinking coats that light up when it's too cold, to signal you should head indoors," Leo said.

Mara rolled her eyes playfully. "And here I thought we were just making self-heating mittens."

The prototypes they developed included gloves that could charge a phone and socks that warmed up when the

---

[20] Japan is at the forefront of sustainable technologies and energy solutions, from advanced solar panels to energy-harvesting roads. Their use of renewable resources has inspired many global research initiatives.

temperature dropped below a certain level. They ran tests in a climate simulation chamber that could mimic extreme conditions.

Leo, wearing the prototype gloves, found that they glowed faintly when he wiggled his fingers. "Is it just me, or do I look like I'm practicing for a sci-fi movie?"[21]

Mara's Surprise Experiment - One weekend, Mara decided to conduct a surprise experiment for Leo—a battery-powered coffee mug that kept his drink warm for hours. She unveiled it on a particularly long research day.

"Leo, meet 'Heat-Mate.'"

Leo's eyes lit up as he took a sip of his perpetually warm coffee. "Mara, this is life-changing. Nobel Prize-worthy, even."

She laughed. "Only if you don't spill it on your research notes this time."[22]

The Scientific Prank - Leo, never one to pass up an opportunity for some lab humour, decided to prank Mara with a harmless setup involving *liquid nitrogen* and marshmallows. He placed the frozen marshmallows in a bowl and offered them as snacks.

Mara picked one up, eyes narrowing. "Is this one of your 'cool' tricks, Leo?"

---

[21] Self-heating textiles often use carbon-based materials that are lightweight yet conduct heat efficiently. By embedding tiny circuits or using smart fabrics, clothing can react to temperature changes and provide warmth or energy.

[22] Battery-powered heated mugs use insulated materials and small rechargeable batteries to maintain the temperature of the contents. Some models use thermoelectric plates that draw minimal power for maximum efficiency.

Before he could answer, she popped it in her mouth and blew out a puff of 'dragon's breath' as the marshmallow sublimated. They both burst out laughing.

"Alright, you win this one," Mara admitted, still chuckling.[23]

Recognition and Future Plans - Their groundbreaking work continued to gain attention. They received an award for *Innovative Materials Research*, which came with funding for a new lab expansion. With their newfound resources, Leo and Mara started exploring the properties of *aerogels*—ultralight, highly porous materials known for their insulating properties.

Leo marvelled at a piece of aerogel, almost weightless in his hand. "This feels like holding a cloud."

Mara added, "A cloud that could insulate a spaceship."[24]

---

[23] When food or objects are dipped in liquid nitrogen, they become extremely cold and can create a vapor effect when consumed, known as 'dragon's breath.' Liquid nitrogen is used in science for its rapid freezing capabilities, but it must be handled with care due to its extremely low temperature.

[24] Aerogels are made by removing the liquid from a gel and replacing it with air, resulting in a material that is 99.8% air. They have extremely low thermal conductivity and are used in space suits and other high-tech applications.

Aerogel Adventure

With funding secured for their lab expansion, Leo and Mara's new project on aerogels captured the attention of the entire university. Aerogels, with their ultralight, yet incredibly strong properties, were poised to be the next game-changer in thermal insulation and aerospace engineering. They spent weeks crafting samples that looked as fragile as spun sugar but could insulate against heat that would fry an egg in seconds.

Leo held up a sample and looked at Mara with wide eyes. "This is practically science's version of magic. The only thing missing is a top hat and a rabbit."

Mara smirked. "You know, if we market this right, we could become the only team in the world to sell 'cloud bricks'."

One day, Leo decided to test the aerogel's strength by standing on a block no thicker than a textbook. It held his weight with ease, but when he tried to demonstrate it to a group of visiting researchers, his shoe slipped, sending him flailing and landing in a pile of bubble wrap stored for fragile shipments.

Dr. Heike, who had come by for a surprise inspection, simply shook his head. "I see you've mastered the art of insulating both materials and your dignity, Leo."[25]

The Disappearing Ink Incident - Leo and Mara took a break one afternoon to conduct an outreach program for local high

---

[25] Aerogels are among the least dense materials known, composed of up to 99.8% air. Despite their lightness, they have excellent thermal resistance and can withstand loads many times their weight.

school students. As part of the presentation, Leo prepared a segment on the properties of *thermochromic materials*— substances that change colour with temperature.

For added fun, Leo wrote "Welcome Future Scientists!" on a whiteboard using disappearing ink made from thermochromic pigments. He had timed it perfectly so that, just as he said, "And this is what real science can do," the message vanished as the room's temperature increased.

However, what Leo hadn't counted on was the ink's tendency to reappear when cooled. Hours later, after they had packed up and left, the janitor found an unexpected message on the board that read, "We know what you did!"—a relic of Leo's demonstration.

The next day, Leo had some explaining to do. "It's educational and a little eerie, which was... totally intentional," he said, trying to keep a straight face as Dr. Heike inspected the board.[26]

The Great Coffee Experiment - Back in the lab, Leo decided to use their latest aerogel samples in a fun, albeit self-serving experiment. He built an insulated coffee cup lined with a thin sheet of aerogel to test how long it could keep his coffee hot.

"Science is supposed to improve lives, and today it's improving mine," he declared as he poured in the coffee.

Hours later, when Mara arrived, she noticed Leo sipping contentedly. "How long has that been warm?" she asked, intrigued.

---

[26] Thermochromic materials change colour based on temperature due to molecular structure changes. They are used in applications like colour-changing mugs, mood rings, and even temperature-sensitive labels.

"About six hours," he said, grinning. "It's like a thermos from the future."

Curiosity piqued, Mara took a sip and yelped. "Leo, it's still boiling! You might've just invented *forever coffee.*"[27]

An Unexpected Collaboration - Word of Leo and Mara's work reached a startup focused on sustainable housing. They proposed collaborating on using aerogels in eco-friendly home insulation. Leo and Mara were excited at the thought of seeing their lab work transition into something tangible.

However, during the first demo at the startup's site, things took a humorous turn when an overly enthusiastic intern decided to test the material's insulation properties by laying on it like a mattress. The aerogel held, but the intern popped up with a static-charged hairdo that defied gravity.

Leo couldn't help but laugh. "Looks like we've found an unexpected use: *anti-gravity hair styling.*"[28]

Mara's Birthday Surprise - Leo, ever the scientist with a flair for theatrics, decided to surprise Mara on her birthday by combining science with celebration. He rigged up a series of balloons with small piezoelectric charges that would pop when clapped near. Each pop released a burst of non-toxic, scented glitter made from biodegradable materials.
The effect was dazzling, though one balloon malfunctioned and sent a plume of glitter right at Leo's face, making him look like a disco ball for the rest of the evening.

---

[27] Aerogel's remarkable insulating properties stem from its nanoporous structure, which minimizes heat conduction, making it ideal for thermal insulators in both consumer and industrial products.

[28] Because of their high surface area and extremely low density, aerogels can accumulate static electricity, especially when manipulated or rubbed. This property is a side effect of their lightweight and porous structure.

Mara, laughing, said, "Leo, I appreciate the thought, but next time, let's keep the glitter contained to the decorations, not the scientist."[29]

The Proposal Redux - Leo's earlier ring invention had become the talk of the research community, so much so that a tech magazine wanted to feature it. To add a twist for the article's photo shoot, Leo and Mara decided to experiment with *electroluminescent* materials to create a glowing edge on the ring. This required delicate wiring and a thin film of phosphor-coated material that emitted light when electrically stimulated.

During the test, Leo accidentally reversed the connections, and instead of a gentle glow, the ring emitted a dazzling flash that left both of them blinking.

"Well, now I know what it feels like to be in a camera flash," Mara said, rubbing her eyes.

Leo winced. "At least we know it works... a bit too well."[30]

Their Most Challenging Experiment Yet - The culmination of their work on aerogels and energy-harvesting tech led them to explore *aerogel solar cells*. These ultralight cells had the potential to be deployed in high-altitude balloons to gather solar energy where traditional solar panels were too heavy.

During one test launch, the balloon carrying the aerogel panels snagged on a tree, prompting Leo to climb up to free it. As he hung awkwardly from a branch, Mara couldn't resist shouting,

---

[29] Piezoelectric materials generate an electric charge in response to mechanical stress. When integrated with small devices, they can be used for creative effects, like popping balloons or generating small sparks.

[30] Electroluminescent materials emit light when an electric current passes through them. They are used in display panels, flexible screens, and light-up fashion items, adding an extra dimension to everyday products.

"Leo, are you testing the tensile strength of the branches or just showing off your climbing skills?"

Leo grinned down at her. "A little bit of both."[31]

Recognition and Future Outlook - Leo and Mara's paper on the practical applications of aerogel technology was published in a major scientific journal, earning them invitations to speak at conferences around the world. They became known not only for their groundbreaking research but for the infectious, humorous approach they brought to science.

At their last talk before the summer break, Leo wrapped up with, "Remember, science isn't just about discovery—it's about laughter, glitter mishaps, and sometimes turning your lab partner into a disco ball."

The audience erupted into laughter, and Mara, standing beside him, whispered, "Here's to our next big adventure."

Leo smiled. "And may it be just as full of science—and glitter."[32]

---

[31] High-altitude solar panels need to be lightweight and capable of withstanding temperature fluctuations and high radiation levels. Aerogel-based solar cells could meet these needs due to their low density and excellent insulating properties.

[32] Public engagement in science, especially when infused with humour, makes complex concepts more approachable. Leo and Mara's light-hearted approach reflects the idea that behind every great scientist is a touch of whimsy.

Elemental Challenge

After the success of their aerogel solar cell project, Leo and Mara decided to experiment with a highly reactive, lightweight material: *liquid metal alloys*. These alloys, often composed of gallium and indium, could conduct electricity and were flexible enough to be used in soft robotics and dynamic circuits. Their latest project aimed to create a circuit that could "self-heal" when broken.

One afternoon, Leo demonstrated the material's self-healing properties by slicing through a wire with a plastic knife. As expected, the cut ends moved back together, reconnecting seamlessly. However, during the demo, he forgot to account for the material's tendency to conduct heat. When he touched it, a sudden warmth startled him, making him jump and knock a beaker off the table.

Mara burst out laughing as Leo grabbed for the beaker mid-air, missing it by a mile. The beaker landed safely on a cushion of lab coats left on a chair. "I guess we can now say your reflexes aren't quite as quick as liquid metal," she teased.[33]

The Magnetic Mayhem - One day, Leo and Mara decided to use *magnetorheological fluids* in an experiment involving adaptive shock absorbers. These fluids become more viscous in the presence of a magnetic field, making them perfect for applications requiring variable resistance.

---

[33] Gallium alloys are used in flexible electronics because they remain liquid at slightly above room temperature but can solidify in cooler environments. Their ability to reform connections makes them ideal for self-healing electrical circuits.

During a demonstration, Leo set up a mini-model car equipped with a vial of the special fluid. When the magnetic field was turned on, the car's suspension stiffened noticeably, to the amazement of their audience. To add a dramatic flair, Leo activated a stronger field, intending to show how stiff the suspension could become.

Unfortunately, the metal clips holding the model car shifted under the magnetic force, causing the car to wobble and eventually leap off the table like it had a mind of its own. It rolled down the aisle and bumped gently into a researcher's chair.

Mara couldn't hold back her giggles. "I guess we just built the first car that's afraid of being showcased."[34]

The Chemiluminescence Craze - One evening, Leo and Mara decided to host a "Science and Snacks" night for their research team. Leo thought it would be fun to demonstrate *chemiluminescence* by mixing luminol with hydrogen peroxide and a catalyst to produce a blue glow. As they prepped the experiment, Leo put on a lab coat that had seen better days.

"Time to light up the lab," he said, pouring the mixture with a flourish.

The room erupted in a beautiful blue glow. However, Mara, noticing Leo's coat reacting to a splash of the mixture, snickered. "Leo, you're glowing!"

Leo looked down to see bright blue spots decorating his coat, making him look like a walking star map. "I guess I've become

---

[34] Magnetorheological fluids change their viscosity in response to magnetic fields. They are often used in car suspension systems, prosthetics, and even earthquake dampers in buildings to reduce vibrations.

one with the experiment," he joked, raising his arms as if to embrace his new luminescent identity.[35]

The Robo-Feline Encounter - Leo's next idea involved creating a *soft robot* with artificial muscles made from a combination of liquid metal and silicone. The concept was to mimic the flexibility of biological tissue, allowing for more realistic movements in robotics. For the first test, Leo designed a robotic cat that could arch its back and purr using tiny actuators.

The test day came, and Mara brought in her real cat, Newton, for some "peer review." Newton watched the robot cat suspiciously as Leo activated it. The robot cat arched its back and let out a mechanical purr.

Newton, not one to be outdone, puffed up and leaped onto the table, swiping at the robot with lightning speed. The robotic cat flailed, its flexible limbs bending at odd angles.

Leo gasped. "We didn't program 'fight or flight' mode!"

Mara scooped up Newton, who was now eyeing the robot victoriously. "I think we found a flaw in the design: it can't handle feline criticism."[36]

A Trip to the Science Expo - Leo and Mara were invited to present their work at a science expo featuring the brightest minds in materials science. Their booth displayed their latest invention—a solar-powered, self-cooling jacket made from

---

[35] Chemiluminescence occurs when a chemical reaction emits light without the need for an external light source. Luminol, often used in forensic science to detect blood traces, emits a blue glow when it reacts with an oxidizing agent like hydrogen peroxide.

[36] Soft robotics use materials like silicone and liquid metals to create robots that can move more naturally and adapt to different environments. They're used in medical devices, wearable technology, and search-and-rescue missions.

phase-change materials that absorbed and released heat based on temperature.

Visitors were wowed as Mara demonstrated the cooling effect by holding a heat lamp near the jacket. Leo, always looking for a laugh, wore the jacket while standing in front of the lamp, striking superhero poses.

"Leo, you know it's not a cape, right?" Mara quipped, shaking her head.

"It's a *cape of comfort*, Mara," Leo replied, tugging on the collar as he basked in the artificial sun.[37]

The Epiphany Moment - As the expo continued, Leo found himself watching other researchers present ideas that sparked new thoughts. A booth showcasing *nanomaterials* for water purification caught his attention. The presenter demonstrated how carbon nanotubes could filter out contaminants at a microscopic level, making even the dirtiest water drinkable.

Leo turned to Mara with a look of inspiration. "What if we combine aerogels with nanotubes for a lightweight, water-purifying backpack? It could be used in disaster zones or remote areas."

Mara's eyes lit up. "Leo, that's actually brilliant. It'd be like a portable, life-saving cloud."[38]

The Award Ceremony Surprise - At the end of the expo, there was an award ceremony for the most innovative research. Leo

---

[37] Phase-change materials absorb or release energy during their transition between solid and liquid states. This property makes them useful in temperature regulation for clothing, building insulation, and electronics.

[38] Carbon nanotubes are cylindrical nanostructures made of carbon atoms, known for their strength and conductivity. They have applications in water purification because of their ability to remove pollutants and pathogens effectively.

and Mara were called up to accept the prize for "Outstanding Practical Application of Materials Science." Leo, in typical fashion, tried to bow and almost knocked over the microphone.

"Well, they did say we made waves," he quipped as he steadied the mic.

Mara took the microphone, smiling. "We're grateful for this recognition and for the journey that's brought us here—filled with laughter, glowing lab coats, and the occasional rogue cat."

The crowd laughed and applauded as Leo and Mara stepped down, their excitement palpable.[39]

---

[39] Science expos and symposiums often lead to collaborative ideas and innovations. Researchers share breakthroughs, which can inspire new projects and foster global cooperation in solving critical challenges.

Liquid Crystal Experiment

Leo's new obsession became *liquid crystal technology*, which fascinated him with its properties of changing light transmission in response to electrical fields. He proposed creating a window that could switch between clear and opaque with a simple voltage change—ideal for energy-efficient buildings.

Their initial tests were promising. One afternoon, Leo decided to give their prototype window a real-world trial in the lab. He stood proudly next to the window, flipping the switch to demonstrate the transition from transparent to opaque.

"It's like having sunglasses for your house!" Leo announced.

But as he flipped the switch back and forth for dramatic effect, the window short-circuited, flickering between clear and opaque like a strobe light. Mara, trying not to laugh, shielded her eyes and teased, "Great, Leo. Now we've invented the world's first dance-club window."[40]

The Cryogenic Surprise - Leo and Mara's lab received a new shipment of liquid nitrogen for their experiments with *cryogenics* and low-temperature materials. Leo, unable to resist the urge to demonstrate the dramatic properties of extreme cold, decided to freeze a rubber ball and show how it shattered like glass when dropped.

---

[40] Liquid crystals are used in displays and smart windows due to their ability to change their alignment and light properties when an electric field is applied. This technology is also behind LCD screens in TVs, computer monitors, and digital clocks.

"Alright, behold the magic of cryogenics!" he said, lifting the frosty ball in front of a small audience of curious students. He dropped the ball, expecting a spectacular break. Instead, it bounced, emitting a dull *thud*.

"Looks like it's resistant to shattering," one student called out, trying to keep a straight face.

Mara smirked. "Leo, did you forget to wait until it was fully frozen?"

Leo looked sheepish. "I guess impatience is my kryptonite."[41]

The Holographic Mishap - For their next research endeavour, Leo and Mara explored *holography* to create 3D images projected in mid-air. Leo was particularly excited to show off a holographic projection of Newton, the cat. The setup was intricate, using interference patterns from laser beams to create the illusion.

"Ready for Newton 2.0?" Leo asked as he activated the projection.

The image of Newton materialized, purring and arching its back. But then the real Newton appeared, spotted his holographic counterpart, and launched into a frenzied attack at the air.

Pandemonium ensued as the real cat batted at the illusion, slipping and landing with a dramatic flop into a box of lab reports. The holographic Newton continued purring as if nothing had happened, its static form unaffected.

---

[41] Liquid nitrogen, at -196°C (-321°F), can freeze objects rapidly. Materials like rubber become brittle when sufficiently cooled, shattering instead of bouncing. This phenomenon helps demonstrate the effect of temperature on molecular motion and elasticity.

"Well, at least one of them came out of this unscathed," Mara said, trying not to burst out laughing.[42]

The Accidental Battery Breakthrough - While testing different *solid-state battery* prototypes, Leo had the idea to use flexible materials that could bend without breaking. He managed to create a battery using graphene and a gel electrolyte, which was thinner than paper and could be folded.

Excited to show Mara, he tucked the battery into his pocket, forgetting it was experimental. Hours later, when he took a seat at a café, the battery made a faint buzzing noise.

Mara glanced over, eyes wide. "Leo, is that your pants or are we getting a call from the future?"

Leo jumped up, frantically patting his pocket. "Oops. It's the experimental battery!"[43]

The Chemistry Cooking Show - Leo and Mara hosted a special lab event called "Chemistry Cooking Night," where they demonstrated food science principles. Leo, channelling his inner TV chef, started by making ice cream using liquid nitrogen.

"Welcome to Leo's Science Kitchen!" he declared, tossing the nitrogen into a bowl of cream and sugar. Clouds of mist poured over the table as the mixture solidified into smooth ice cream.

Mara, following up with a segment on *spherification*, showed the crowd how to create "caviar" made from fruit juice and

---

[42] Holography uses the interference of light waves to create a three-dimensional image. The process involves splitting a laser beam into two paths: one reflects off the object and interferes with the other beam to create a pattern recorded on a medium.

[43] Graphene, a single layer of carbon atoms arranged in a two-dimensional lattice, is renowned for its strength and conductivity. It is often researched for use in advanced batteries due to its potential for faster charging and higher energy capacity.

calcium chloride. Just as she popped a perfectly spherical bead into her mouth, Leo presented his ice cream. But in his excitement, he underestimated the freezing power of liquid nitrogen, causing the spoon to emit a tiny frost puff when he handed it over.

"I think you've redefined brain freeze, Leo," Mara said, pretending to shiver.[44]

The Conference Chaos- Leo and Mara were invited to speak at an international conference on future materials. Their topic: *shape-memory alloys* that could return to their original shape after being deformed when exposed to heat.

Leo demonstrated with a coil that straightened out when dunked in hot water. The audience applauded as he explained how such materials could revolutionize medical devices and robotics.

Suddenly, the coil popped out of the water and landed on the table, rolling toward the edge. Leo lunged to catch it but tripped, ending up with a face-full of conference pamphlets. The audience burst into laughter, while Leo raised his hand in triumph, coil in hand.

"See? It even makes me spring into action!" he said, trying to save face.[45]

The Floating Garden - Inspired by hydroponics and materials science, Leo and Mara worked on creating a "floating garden" using lightweight aerogels that absorbed water and nutrients

---

[44] Liquid nitrogen is often used in molecular gastronomy to rapidly freeze foods, creating a smooth texture. Spherification uses a chemical reaction between sodium alginate and calcium chloride to form a thin, gel-like membrane around liquids.

[45] Shape-memory alloys, such as Nitinol (nickel-titanium), can "remember" their original shape. When deformed and then exposed to heat, they revert to their original form. This property makes them useful in stents, actuators, and adaptive eyewear frames.

but were buoyant enough to float. Their goal was to test this concept as a potential solution for urban farming on rooftops and in water-scarce regions. The initial trial was promising, but they had underestimated the growth speed of certain plants. By the end of the week, the small lab garden resembled an overgrown jungle.

"Leo, I think we accidentally created a *mini Rainforest Café*," Mara said, trying to navigate the vines and leaves.

Leo stood in the middle, looking both bewildered and amused. "At least we know it works... maybe a little too well."[46]

The Proposal 2.0 - Leo decided that after years of playful mishaps and amazing scientific feats, he needed to surprise Mara with another, more heartfelt gesture. This time, he prepared a proposal featuring their mutual passion—science.

In the lab, he designed a ring that glowed softly using an *organic light-emitting diode (OLED)*, powered by an ultra-thin flexible battery. He hid the ring in a solution of clear hydrogel. When Mara walked in, he guided her through a mini experiment to extract the ring, pretending it was a new type of sensory polymer test.

As the gel dissolved, the glowing ring emerged. Mara gasped, eyes sparkling. "Leo, this is... perfect."
Leo, looking slightly nervous, knelt. "It's been an electrifying journey. Will you keep exploring with me?"
Mara laughed, tears in her eyes. "Always, Leo."[47]

---

[46] Aerogels can hold up to 99.8% air by volume and have incredible water absorption properties. Combining them with hydroponics could make for a sustainable and efficient growing method, especially in environments where soil isn't practical.

[47] OLEDs emit light in response to an electric current and are used in displays for phones, TVs, and other screens. They can be flexible and emit various colours, making them suitable for wearables and innovative applications.

The Quantum Comedy

Leo and Mara, buoyed by the success of their floating garden project, shifted their focus to a new frontier: *quantum materials.* The idea of manipulating particles that didn't follow classical rules fascinated Leo, who couldn't resist the challenge of bringing quantum physics into practical materials science.

One afternoon, while experimenting with a material known as *graphene oxide,* which could theoretically demonstrate quantum tunnelling, Leo began to explain its properties to Mara.

"So, imagine this," Leo said, eyes wide with excitement. "Electrons behave like they're teleporting through an energy barrier. They don't go over it; they just... appear on the other side!"

Mara raised an eyebrow. "So, like when I lose my pen, and it mysteriously shows up on the other side of the room?"

"Exactly!" Leo exclaimed. "Quantum physics: making lost-and-found services obsolete."[48]

The Infrared Ink Incident - Leo and Mara's next endeavour was creating *infrared-sensitive inks* that only showed text or images when exposed to specific wavelengths of light. Their aim was to develop a "privacy ink" that could hide important information until revealed under infrared illumination.

---

[48] Quantum tunnelling is a phenomenon where particles move through a barrier that they shouldn't be able to pass according to classical physics. This concept is vital in technologies like tunnel diodes and some forms of flash memory.

After hours of perfecting the formula, Leo was ready to test it. He wrote, "This is a top-secret message," on a piece of paper and invited Mara to read it under an infrared lamp.

Mara held the lamp, but instead of revealing the message, the ink glowed erratically, forming the words: "Hello, Newton was here." Both turned to see Newton pawing at a container of infrared dye.

Leo sighed. "Great, now even the cat is a cryptographer."

Mara chuckled. "Newton's probably developing his own security protocols."[49]

The Great Robot Dance-Off - Leo and Mara were invited to showcase their work at a tech fair, where they decided to demonstrate their soft robotics technology. They programmed their robots to move fluidly, using artificial muscles made from electroactive polymers.

To grab attention, Leo suggested a "robot dance-off." They choreographed the robots to perform synchronized moves to an upbeat song. Everything went perfectly until a miscalculation in coding caused one robot to spin continuously, taking out another bot with it.

The crowd erupted in laughter as Mara quickly recovered the rogue robot. "Looks like we programmed a breakdancer."

Leo, unfazed, added, "Well, it's called the robot shuffle for a reason."[50]

---

[49] Infrared inks use pigments that absorb and reflect infrared light, making them visible only under specific conditions. This technology is used for anti-counterfeiting measures and covert communications.

[50] Electroactive polymers can change shape or size when exposed to electrical stimuli, making them ideal for soft robotics. These materials are useful in creating artificial muscles for prosthetics and robots that require flexible, human-like movement.

The Bioluminescence Gala - For a change of pace, Leo and Mara attended a gala celebrating *bioluminescence research.* Guests were encouraged to dress in outfits inspired by glowing sea creatures. Leo, ever the innovator, incorporated actual bioluminescent proteins into the fabric of his jacket, which emitted a soft green glow.

"Nice touch, Leo," Mara said, wearing a dress decorated with LED threads that mimicked jellyfish tentacles. "We're like the stars of a deep-sea documentary."

However, midway through the event, Leo's jacket started glowing more intensely, drawing attention from everyone around. Mara noticed Newton's collar had been turned into an impromptu "battery," causing the glow to amplify.

"Looks like Newton's stealing the show," Leo whispered as the cat proudly paraded around.[51]

The Surprise Lightning Exhibit - Mara suggested they try something adventurous: creating a *mini lightning generator* using a Van de Graaff generator and specialized electrodes. The plan was to showcase it as an educational piece in their community science centre.

The setup worked beautifully—arcs of electricity danced between electrodes, crackling with energy. Leo, however, couldn't resist a dramatic flourish, stretching out his hand to the generator. Sparks jumped to his fingers, causing his hair to stand up in gravity-defying spikes.

A group of children nearby laughed. One called out, "Mr. Einstein's got competition!"

---

[51] Bioluminescence is the production of light by living organisms, seen in species like fireflies and certain jellyfish. This light is produced through a chemical reaction involving the enzyme luciferase and the substrate luciferin.

Mara, trying not to laugh too hard, snapped a photo. "Leo, I think you've officially charged up the crowd."[52]

A Night Under the Stars - One night, Leo and Mara decided to take a break from the lab and camp under the stars, taking a telescope to observe the night sky. Leo, being Leo, brought along a spectrometer to analyse starlight and determine their chemical composition.

"Did you know that most of what we see up there is hydrogen and helium?" Leo said, pointing at a particularly bright star.

Mara nodded. "And yet, humans spend their entire lives chasing other elements. Iron, gold..."

Leo grinned. "Yeah, we should probably just tell them that the real treasure is *stardust*. Literally, we're made of the same stuff as stars."

As they continued stargazing, Mara leaned over. "Leo, ever think about what's next?"

Leo, pretending to be lost in thought, answered, "Yeah, how do we convince Newton not to climb the telescope?"

Newton was, indeed, perched halfway up the tripod, gazing at the night sky as if he had scientific thoughts of his own.[53]

The Final Project Proposal - Back at the lab, Leo and Mara started working on a groundbreaking project: creating a material capable of capturing carbon dioxide and converting

---

[52] A Van de Graaff generator can create high-voltage electricity, which causes static electricity effects, like hair standing on end. The charged dome attracts electrons from the hair, making the hairs repel each other due to similar electric charges.

[53] Spectroscopy involves analysing the light emitted by stars to identify the elements present. Each element absorbs and emits light at specific wavelengths, creating a unique spectrum that can be used as a "fingerprint" for identifying its presence.

it into a useful substance. They decided to integrate *metal-organic frameworks (MOFs)* with a catalytic system.

Leo proposed, "What if we can turn CO2 into something practical, like biodegradable plastic?"

Mara's eyes lit up. "That's the kind of world-changing idea that could inspire generations."

As they worked on their first prototype, Leo accidentally spilled a catalyst, causing a chain reaction that turned a section of their bench into a biodegradable, jelly-like substance.

Mara laughed. "Well, it's not exactly plastic, but it's a start!"[54]
The Award-Winning Conclusion - Leo and Mara's work on their CO2 conversion project caught the attention of several environmental organizations. They were invited to present their findings at a global sustainability summit. The day they presented, Leo stumbled over a wire and nearly sent their model flying, but Mara caught it just in time, earning an amused cheer from the audience.

The presentation ended with Leo saying, "Remember, sometimes the biggest discoveries come from happy accidents—or in our case, cat-related mishaps."

Their humour, combined with their innovative project, earned them a standing ovation. As they left the stage, Mara whispered, "So, what's our next adventure?"
Leo smiled, "Saving the world, one chaotic experiment at a time."[55]

---

[54] MOFs are porous materials made up of metal ions and organic linkers that can store gases, including CO2. Researchers are exploring their potential in carbon capture and conversion technologies.

[55] Carbon capture and utilization is a field focused on turning CO2 emissions into useful products, such as fuels, chemicals, and building mate

Piezoelectric Predicament

Leo and Mara embarked on their next innovative challenge: developing a *piezoelectric material* that could generate electricity from mechanical stress. Their vision was to integrate these materials into sidewalks, so walking could help power streetlights.

Leo explained their plan to a group of interested students in the lab. "So, when you step on these panels, the pressure will create a small electrical charge. Basically, we're turning footsteps into tiny power stations."

One eager student stepped onto the prototype panel a bit too enthusiastically. The panel emitted a loud *pop*, causing everyone to jump. Mara, covering her laugh, said, "Looks like we just discovered the world's first sound-powered electricity."

Leo winced and held up the scorched remains of the panel. "Step by step, literally."[56]

The Graphene Skateboard - Leo's love for graphene hadn't waned. He convinced Mara to help him build a *graphene composite skateboard* to demonstrate the material's strength and flexibility. Their first test ride in the university courtyard attracted attention. Leo, decked out in full safety gear, pushed off while Mara filmed.

---

[56] Piezoelectric materials, like quartz and certain ceramics, generate an electric charge in response to mechanical stress. They are used in various applications, including sensors, lighters, and even energy-harvesting floors.

Halfway through a smooth glide, Leo hit a small rock. The skateboard, true to graphene's properties, flexed slightly and kept going, but Leo's balance failed him. He landed with an ungraceful tumble, the skateboard continuing on its merry way.

Students nearby applauded. "Nice trick!" someone shouted.

Leo stood up, brushing off his helmet. "Next time, graphene-powered brakes," he muttered.[57]

The Electrochromic Prank - To test their theories on *electrochromic windows* that change colour with applied voltage, Leo and Mara set up a window in their lab. By adjusting the voltage, they could shift the window from clear to various shades of blue. They discovered an unexpected side effect—static electricity built up when the voltage changed too quickly.

One day, Leo mischievously altered the settings when Mara walked past. The window flashed blue, and Mara's hair stood on end like a dandelion puff. She stopped, wide-eyed, as Leo doubled over in laughter.

"You're lucky I'm not holding a beaker!" Mara warned, though she couldn't hide her smile.[58]

The Hydrogen Balloon Bonanza - Leo and Mara wanted to make their chemistry classes more engaging, so they organized a demonstration with *hydrogen balloons*. The idea

---

[57] Graphene is a single layer of carbon atoms arranged in a hexagonal lattice. It's incredibly strong—around 200 times stronger than steel—and highly flexible. Research into composite materials aims to harness these properties for lighter, more durable products.

[58] Electrochromic materials change their optical properties (like colour) when an electrical current is applied. These materials are used in smart windows for energy-efficient buildings, providing privacy and reducing heating and cooling costs.

was to show how hydrogen, being highly flammable, reacts with oxygen to create water in a controlled mini-explosion.

Leo stood at the front, match in hand, as Mara narrated. "When we ignite this balloon, hydrogen reacts with oxygen, releasing energy in the form of heat and light."

Leo touched the match to the balloon, which popped with a resounding *boom*, startling everyone. The room was silent for a moment before Mara broke it. "And that's why we don't use hydrogen in party balloons."

One of the students called out, "Best science class ever!"[59]

The Robot That Went Rogue - Leo and Mara's research on soft robotics had evolved into building a prototype of a *self-healing robot* using materials infused with microcapsules of healing agents. One afternoon, Leo gave their robot—nicknamed "Patch"—its first task: walking across a makeshift course lined with small nails to simulate damage.

Patch started well, moving with an awkward but determined gait. Then it stepped on a nail, causing a tiny tear. The robot paused, secreted a gel from its microcapsules, and sealed the wound. The crowd watching gasped in awe.

Suddenly, Patch veered off course and bumped into a table leg repeatedly.

Mara sighed. "Patch has a mind of its own."

Leo shrugged. "It's not self-aware, just stubborn."[60]

---

[59] Hydrogen is the lightest and most abundant element in the universe. Its flammability is due to its tendency to react with oxygen, forming water and releasing energy—a principle that's harnessed in rocket fuel.

[60] Self-healing materials contain microcapsules filled with substances like polymers or resins that are released when damage occurs. The reaction seals the damage, extending the

The Power of Solar Paint - Inspired by sustainability, Leo and Mara researched *solar paint*, a type of coating that could convert sunlight into electricity. They painted a small section of their lab wall and hooked it up to a basic circuit with an LED light.

"It's not about how much light we create," Leo said, "but about proving it's possible."

The LED flickered on, dimly at first but brightening as the sun reached its peak. Excited, Mara high-fived Leo. Newton, not to be left out, jumped onto the table and swatted the LED. The connection broke, and the light went out.

"Well, at least the cat is off the grid now," Mara joked.[61]

The Silliest Science Symposium - Leo and Mara were invited to present at a "Silliest Science Symposium," where researchers showcased their most amusing experiments. They decided to feature their *bio-inspired dancing robots*, which mimicked the movements of jellyfish using flexible actuators.

The show went well until a synchronization glitch caused the robots to move erratically, bumping into each other and forming a conga line. The audience roared with laughter as one robot, the "leader," spun in circles, setting off the others like dominoes.

Mara whispered, "I guess it's less a dance and more a robot mosh pit."

---

material's life. This technology is being developed for use in electronics, construction, and robotics.

[61] Solar paint uses nanoparticles known as quantum dots or perovskites to convert sunlight into electricity. While still in the research phase, this paint could potentially transform any surface into a source of renewable energy.

Leo shrugged, "At least they're enthusiastic."[62]

The Surprise Proposal Upgrade - After their hilarious symposium success, Leo decided to upgrade his original proposal ring with a scientific twist. He embedded a tiny *pressure sensor* into the ring so that when Mara wore it, the sensor would glow when it detected a change in pressure, like when they held hands.

During a walk through the campus park, Leo presented the ring. "Now, our connection lights up—literally."

Mara slid the ring on, and it emitted a soft, golden glow as she intertwined her fingers with Leo's. "I'm glowing," she whispered, tears in her eyes.

"Actually, that's a low-voltage circuit activating the—" Leo started, but Mara cut him off with a kiss.[63]

The Accidental Carbon Capture - Leo and Mara's CO_2 conversion research led them to a breakthrough when they combined MOFs with a catalyst made from recycled coffee grounds. The result was a sponge-like material that efficiently absorbed CO_2 and converted it into harmless carbonates.

As a side project, Leo brought a sample home to see if it could handle the output of Newton's play-induced panting. The next morning, he found Mara grinning.

"Look, Leo, the air filter worked so well it might've absorbed *our* CO_2 too. I haven't yawned once!"

---

[62] Bio-inspired robotics use design principles observed in nature, such as the fluid motion of jellyfish, to create more efficient and adaptive machines. These designs are crucial in underwater exploration and soft robotics.

[63] Pressure sensors convert force or pressure into an electrical signal. These are used in various applications, from smartphones to wearable tech, and can be miniaturized for innovative uses, like Leo's glowing ring.

Leo chuckled. "Mara, we might have created the world's first sleep-enhancing CO2 filter."[64]

The End (Or Just the Beginning)

Leo and Mara's journey continued, marked by new discoveries, whimsical failures, and the occasional cat intervention. Their unique blend of humour, relentless curiosity, and love for science made every day an adventure worth living.

---

[64] MOFs' high surface area makes them excellent for absorbing gases like CO2. Combining them with catalysts can help convert CO2 into other chemicals, potentially helping reduce greenhouse gases.

The Misbehaving Magnet Experiment

Leo and Mara had grown accustomed to the unexpected when conducting experiments, but their project on *magnetic shape-memory alloys* introduced a new level of unpredictability. These materials could change shape when exposed to a magnetic field and were being studied for potential use in actuators and sensors.

During a demonstration in their home lab, Leo placed a sample between two magnets. As he cranked up the electromagnet, the alloy bent and twisted as if performing an interpretive dance.

"It's like it's alive!" Mara exclaimed, barely stifling a laugh as the alloy wriggled out of the clamps and fell to the floor.

Newton pounced on it immediately, batting it across the room as Leo and Mara chased him in circles. Leo tripped over an extension cord, landing on the floor just as Newton released the alloy into the recycling bin.

Mara clapped her hands. "Well, there's one way to recycle experiments."[65]

The Liquid Nitrogen Ice Cream Stand - Leo and Mara's next venture was more crowd-pleasing: making *liquid nitrogen ice cream* at the local science fair. They set up their booth, complete with goggles, lab coats, and a tank of liquid nitrogen.

---

[65] Magnetic shape-memory alloys are materials that can change their shape when exposed to a magnetic field. They're being explored for use in precision actuators and microelectromechanical systems (MEMS).

"Step right up and get the fastest ice cream in the universe!" Leo announced to a gathering crowd. He poured the creamy mixture into a bowl, added a dash of liquid nitrogen, and stirred vigorously as fog billowed up dramatically.

A young child approached and tugged at Mara's coat. "Is that safe?"

Mara winked. "Only if you don't touch it until it stops smoking. Unless you want a tongue-cicle."

The demonstration was a hit, with Leo accidentally freezing a spoon to the table and Mara blowing vapor rings with the leftover nitrogen. Their unconventional ice cream flavours, like "Periodic Table Pistachio" and "Entropy Espresso," sold out within hours.[66]

The Quantum Cat Conundrum - Leo and Mara were invited to give a guest lecture on *quantum mechanics* at the university, which they titled "Schrödinger's Cat Walks Into a Bar." They filled the talk with humorous anecdotes and simplified explanations.

"Imagine this," Leo said, gesturing dramatically. "Newton's in a box. But according to quantum theory, he's both purring and plotting to knock it over at the same time until you open it."

The audience laughed as Newton, lounging at the back of the room, sneezed in what could only be interpreted as disapproval.

---

[66] Liquid nitrogen has a boiling point of -196°C (-321°F), allowing it to freeze substances quickly. This rapid freezing creates tiny ice crystals, resulting in smoother ice cream.

Mara chimed in, "And the question isn't just if he's in the box, but how many times we'll have to pick it up after he's knocked it over."[67]

The Great Battery Battle - With the rise of *solid-state batteries* becoming a hot topic, Leo and Mara decided to prototype a safer and more efficient version for portable electronics. The duo mixed and matched various lithium compounds with an electrolyte gel made from a secret ingredient: recycled kelp.

"What's the kelp for?" asked one of their colleagues who dropped by to check out their project.

"Eco-friendliness," Leo replied, grinning. "Also, it's easier to pronounce than tetrafluoroethane."

Their first test was a success—until they tried charging the battery too quickly. The cell emitted a loud *pop*, spewing a puff of greenish smoke.

Leo fanned the air. "On the bright side, we've just invented seaweed-scented smoke detectors."

Mara added, "Patent pending."[68]

The Solar Car Fail (and Victory) - Leo and Mara collaborated on a solar-powered car for a university competition. The project involved attaching solar panels with flexible, transparent solar cells that could contour to the shape of the car. Their prototype, dubbed "Photon Flinger," looked sleek and futuristic.

---

[67] Schrödinger's cat is a thought experiment illustrating the concept of superposition in quantum mechanics. It shows that a particle can exist in multiple states simultaneously until it's observed.

[68] Solid-state batteries use a solid electrolyte, which can make them safer and offer higher energy density than conventional liquid electrolyte batteries. Research is ongoing to make these technologies more viable and environmentally friendly.

Race day arrived, and Leo suited up in a comically oversized helmet. As Mara counted down, he pressed the start button, only for the car to emit a pathetic *whirr* and stay motionless. A stray pigeon took off from the nearby bench as if mocking them.

Mara tilted her head. "I think your car's in nap mode."

After a quick check, they discovered a tiny connector had come loose. With a flick and a push, "Photon Flinger" finally zipped across the track, overtaking the competition.

The crowd cheered, and Leo waved, helmet bobbing. "Looks like I need a pit crew next time."[69]

The Day Mara Built a Miniature Black Hole (Almost) - During one of their astrophysics side projects, Mara playfully proposed building a simulation of a black hole using a rotating light array and carefully placed mirrors to demonstrate gravitational lensing.

Leo, always up for the challenge, helped assemble the contraption. As they tested it, the setup created a perfect ring of light, resembling an *Einstein ring.*

Mara joked, "Now we just need Stephen Hawking's hologram to narrate this."

Leo adjusted the angle, making the ring wobble and distort. "If this collapses the room, I blame Newton," he said, pointing to the cat who was batting at one of the cables.[70]

---

[69] Solar panels used in vehicles are typically made from materials like monocrystalline silicon, known for their high efficiency. New research into perovskite solar cells holds promise for making solar panels lighter and more adaptable to curved surfaces.

[70] Gravitational lensing occurs when light from a distant source is bent around a massive object, such as a black hole, due to its gravitational pull. This phenomenon has been used by astronomers to study galaxies and dark matter.

A Proposal of Galactic Proportions - As Leo and Mara wrapped up their final research trip before transitioning to their next chapter in life, Leo had a surprise planned. They visited an observatory to stargaze, surrounded by the hum of telescopes and distant constellations. Leo fumbled nervously with a small device in his pocket—a laser pointer modified to project words onto the sky.

When Mara pointed out a particularly bright star, Leo activated the pointer. Words appeared against the night backdrop: "Will you explore the universe with me, Mara?"

Mara gasped, eyes wide with delight. "Leo, you science-show-off."

They sealed the moment with a kiss under the sky, filled with the light of stars, nebulas, and endless possibilities.[71]

---

[71] Laser pointers, when modified with lenses and appropriate focus, can project patterns or words over large distances. Observatories use lasers for calibrating and aligning telescopes, particularly in adaptive optics systems.

Antigravity Shoes Fiasco

Leo, inspired by recent research on *negative mass* materials and gravitational manipulation theories, decided to create a pair of "antigravity" shoes. The idea was simple: integrate powerful electromagnets to reduce a person's weight by pushing against the Earth's magnetic field.

He presented the prototype to Mara with a grin. "Behold! Shoes that will make you feel like you're walking on the moon."

Mara slipped on the shoes while Leo activated them. Initially, everything seemed fine. But as she took a few cautious steps, one shoe began to levitate slightly higher than the other. Mara flailed, spinning in a slow pirouette.

"Leo!" she shouted, laughter in her voice. "How do I stop this waltz before I end up in orbit?"

Leo scrambled to deactivate the shoes, his face red from suppressed laughter. "I guess I didn't account for rotational torque."

Newton, watching from the side, seemed to roll his eyes before swatting at the stray shoelaces.[72]

The Chemical Cocktail Party - Leo and Mara hosted a "Science and Silliness" cocktail party, where drinks were inspired by chemical reactions. The "Ethanol Explosion" featured a safe, bubbling concoction of dry ice and fruit juice, while "Glow-

---

[72] Negative mass is a theoretical concept where an object would move opposite to the direction of force applied to it. While real-world antigravity remains science fiction, researchers study materials that can exhibit repelling properties through magnetic fields.

tini" was a bright green drink that glowed under blacklight thanks to quinine in tonic water.

Mara held up a glowing glass, winking at Leo. "Who knew a bit of quinine could make us feel like we're in a sci-fi movie?"

A guest took a sip and gasped. "It tastes like neon!"

Leo laughed. "Well, technically, neon doesn't taste like anything—it's an inert gas. But we can pretend."

Newton, not to be left out, padded around with a collar that reflected the blacklight, looking like a cosmic creature.[73]

The Superhydrophobic Bench Incident - Leo's latest experiment involved creating a *superhydrophobic* surface that repelled water like a duck's feathers but on steroids. He coated a bench outside their lab with a specially formulated nano-coating and invited Mara to test it out.

"It's supposed to be self-cleaning and perfect for outdoor furniture," Leo explained. Mara eyed the bench, sat down carefully, and marvelled as a sudden drizzle left her dry while water droplets skittered off like tiny dancers.

Just then, a fellow student, oblivious to the experiment, ran over and tried to sit beside her. He slid off in slow motion, landing on the grass with a look of pure confusion.

Leo stifled a chuckle as Mara called out, "It's not you—it's physics!"[74]

---

[73] Quinine, found in tonic water, fluoresces under ultraviolet light, making it glow. It's also historically used to treat malaria, and in safe doses, it's what gives tonic water its characteristic bitter flavour.

[74] Superhydrophobic surfaces are created by applying a micro- or nanostructured coating that significantly reduces the adhesion of water. These surfaces mimic the lotus leaf effect, where water forms droplets and rolls off, taking dirt with it.

The Attempted AI Assistant Gone Rogue - Leo and Mara's venture into artificial intelligence took an unexpected turn when they developed "ALFIE" (Algorithmic Lab Friend and Inquiry Engine). The idea was for ALFIE to assist with calculations, suggest hypotheses, and provide comic relief.

On the day of the first test, ALFIE's voice boomed from the speakers. "Welcome, esteemed scientists! What can I do today? Also, did you know bananas are berries but strawberries are not?"

Leo raised an eyebrow. "Great start, ALFIE. Calculate the tensile strength of our latest graphene composite."

ALFIE hummed and responded, "Sure! But first, a joke: Why was the math book sad? Because it had too many problems."

Mara rolled her eyes, suppressing a giggle. "ALFIE, calculations first, jokes later."

After a few adjustments, ALFIE behaved better, though it occasionally interrupted with, "You know, I'd be better at this if I had arms," sparking debates among the lab team about the necessity of AI limbs.[75]

The Bioengineered Plant That Did Too Well - Leo, inspired by the potential of *genetically modified organisms*, experimented with splicing a plant's DNA to create a super-efficient photosynthesis process. The result? A small, potted plant named "Verdie" that produced twice the usual oxygen and glowed slightly in the dark due to modified chloroplasts.

---

[75] AI systems use complex algorithms, often based on machine learning, to process information, learn from data, and make predictions. While current AI lacks self-awareness, it can mimic conversational and decision-making abilities through programmed logic.

Mara walked into the lab one evening and stopped short. "Leo, why is the plant... humming?"

Leo peered at Verdie, which emitted a low, soothing sound. "Oh, that's just the result of its faster metabolic cycle. Completely normal—probably."

Verdie's leaves waved, almost as if acknowledging them. Newton pounced on it, and Verdie reflexively released a puff of harmless bioluminescent mist.

Leo sighed. "Note to self: check for unintended side effects."[76]

The Cold Fusion Commotion - Leo and Mara's next big project was a theoretical attempt at *cold fusion*, a process that, if successful, could revolutionize energy production by creating limitless power from water. The experiment involved deuterium and palladium electrodes submerged in a pressurized chamber.

As they waited for any sign of an energy spike, Mara joked, "If this works, I'm naming our first kid 'Deuteria.'"

Leo laughed nervously. "And if it doesn't work, Newton gets to keep the lab."

The gauges flickered, but instead of producing power, the setup emitted a strange "blip" and a gentle stream of bubbles. Newton padded over and tried to bat at them.

Leo scratched his head. "Looks like we've invented the world's most expensive bubble machine."[77]

---

[76] Genetic engineering can modify plants for various traits, such as increased photosynthetic efficiency or resistance to pests. This technology has the potential to improve agricultural productivity and reduce resource use.

[77] Cold fusion refers to a hypothesized nuclear reaction that would occur at room temperature, unlike the extreme conditions needed for hot fusion. Despite claims and research, achieving practical cold fusion remains unproven.

The Impromptu Workshop at the Beach - Leo and Mara took a much-needed break at a nearby beach but couldn't resist conducting a quick experiment with *solar evaporation*. Using simple materials, they built a solar still that purified seawater.

"Free science lesson, anyone?" Leo shouted, drawing in beachgoers who watched as salty water condensed into drinkable drops.

A kid asked, "What if you did this with soda?"

Mara chuckled. "You'd end up with sweet, sticky water—and very confused ants."

The experiment was a hit, and even Newton, donning a tiny sun hat, seemed to approve.[78]

The Gravity-Powered Desk Organizer - Back at the university, Leo worked on a device that could use *gravity* to power small tasks, like rolling marbles to keep papers tidy or light up tiny LEDs as weights shifted.

Mara raised an eyebrow. "A paperweight that works out? Impressive."

Newton, ever curious, nudged the marble track, sending the balls spiralling around, triggering lights and levers. One marble shot out and landed in Leo's coffee cup, splashing him.

Leo wiped his face, sighing. "Newton: zero respect for gravity."[79]

---

[78] A solar still uses the sun's heat to evaporate water, leaving salts and impurities behind. The vapor then condenses on a cool surface and drips into a container. This method is simple and sustainable for producing freshwater.

[79] Gravity-powered devices are explored in renewable energy systems, such as gravity batteries, which store energy by lifting weights and release it as the weights descend, driving turbines.

The Proposal Upgrade 2.0 - Leo, always thinking one step ahead, revealed his second proposal surprise. He had embedded *smart rings* with tiny sensors that lit up when they were near each other, symbolizing their connection.

When Leo gave Mara her ring, it glowed softly as he slid it onto her finger. "This is so that no matter where we are, we'll always light up each other's world."

Mara's eyes sparkled with tears. "You and your inventions."

Newton meowed, as if to say, "Where's mine?"

Leo glanced at Mara. "Should we start working on a glowing cat collar next?"

Invisible Ink Incident

Leo's next experiment involved *photochromic compounds*, materials that change their appearance when exposed to UV light. His idea was to create a secret message that would only reveal itself outdoors under the sun. With a mischievous smile, he wrote "Free Cookies Inside" on the door of their lab using this special ink.

Mara discovered it when a confused professor stopped by, squinting at the door and muttering, "I don't remember approving a bake sale."

She raised an eyebrow at Leo. "Do we actually have cookies, or is this an experiment in disappointment?"

Leo shrugged. "It's a lesson in perception and UV light. And no, there aren't any cookies."

The next day, they found Newton staring at a line of students waiting outside the lab, all with hopeful looks. One of the students muttered, "This better not be a chemistry joke."[80]

The Unexpected Superconducting Donut - Leo and Mara took on the challenge of building a small *superconducting maglev* model using liquid nitrogen. The base track was cooled to near absolute zero, allowing a superconducting material to levitate due to the Meissner effect. Leo fashioned the levitating piece into the shape of a donut. Mara eyed it sceptically.

---

[80] Photochromic compounds change their chemical structure when exposed to UV light, which alters their colour. This is the principle behind transition lenses in glasses, which darken in the sunlight.

"You made a floating donut and didn't think to add sprinkles?"

As the donut floated smoothly around the track, Newton, ever the curious cat, leapt up to paw at it, resulting in an elegant, mid-air twist before it settled back on the track. The entire lab burst into laughter as Newton meowed in surprise.

Leo said, "Who knew we'd invent the world's first cat-proof flying breakfast?"[81]

The Synchronized Robot Tango - Leo's newfound fascination with *robotic choreography* led to an ambitious project: programming a set of small robots to perform a synchronized dance. Each robot was equipped with lightweight, flexible materials that responded to simple commands.

Leo stood back as the music played, and the robots spun, shuffled, and twirled in unison. One, however, seemed to interpret "spin" as "spin wildly" and pirouetted off the table.

Mara picked it up, shaking her head with a smile. "Looks like this one's auditioning for a solo."

A student observing the experiment joked, "Is this the new era of performance art?"

Leo beamed. "It's science meeting culture. Also, I'm teaching robots how to express themselves, one misstep at a time."[82]

The World's First Noise-Cancelling Coffee Cup - Fed up with the constant din of the lab, Leo designed an experimental

---

[81] Superconductors exhibit zero electrical resistance and expel magnetic fields when cooled below a certain temperature, allowing them to levitate above magnets. The Meissner effect is what makes this possible and is crucial in maglev train technology.

[82] Robotic choreography involves programming motion paths and using sensors to adjust movement in real-time. Advances in lightweight, shape-memory alloys and adaptive materials allow robots to execute complex manoeuvres that mimic natural motion.

noise-cancelling coffee cup. Using principles similar to those in noise-cancelling headphones, the cup emitted counter-frequency sound waves to create a zone of silence around the drinker.

Mara tested it during a particularly loud morning when Newton was practicing his new skill of batting a marble across the floor. She lifted the cup and took a sip, her eyes widening.

"This is amazing! It's like drinking in a vacuum. I can't hear anything," she said.

Just then, Leo's voice crackled through the room's intercom. "Reminder: don't drink in space, where you'd need an actual vacuum suit."

Mara rolled her eyes and held up the cup as if in toast. "Cheers to quiet breakthroughs."[83]

The Multitasking Desk - Leo's latest creation was an AI-powered *multitasking desk* that organized schedules, shifted between different heights, and even dispensed stationery on command. It had built-in sensors that detected who was at the desk and adjusted settings accordingly.

Mara tested it, saying, "Pen, please." A compartment opened smoothly, and a pen popped out. But when Leo leaned over and said, "Stapler, please," it flung a stapler out with the enthusiasm of a catapult.

Mara caught it mid-air, laughing. "Leo, did you set it to aggressive mode?"
He checked the interface, wincing. "It seems to have an 'over-eager assistant' setting. We'll tone that down."

---

[83] Noise-cancelling technology works by detecting ambient noise with a microphone and producing sound waves that are 180 degrees out of phase with the noise. These sound waves cancel out the original noise through destructive interference.

Newton settled on the desk, triggering it to dispense a paperclip. He promptly batted it onto the floor, repeating the process until a pile formed.

Leo sighed. "Great, we've just made a cat-friendly vending machine."[84]

The Failed Space Balloon Experiment - For their next ambitious project, Leo and Mara tried to send a *weather balloon* into the stratosphere with a mini-camera attached. They carefully prepped a payload containing temperature sensors and a handwritten note that read, "Hello from Leo and Mara!"

As the balloon lifted, Newton sat by the window, eyes wide with fascination. The camera feed showed a stunning view of the Earth below—until a sudden gust of wind sent the balloon spiralling and the feed cut off.

Mara sighed. "Well, it's not lost. It's just... somewhere else."

A week later, they received an email from a farmer in France, who found their balloon deflated in his field. Attached was a selfie with the note and a caption: "Bonjour from the countryside!"

Leo grinned. "Looks like our project turned into international outreach."[85]

The Portable Water Purifier - Leo and Mara's next breakthrough was a portable water purifier that used a

---

[84] Modern smart desks incorporate IoT technology and AI to optimize user posture, improve productivity, and respond to voice commands. While current models don't dispense objects, robotics advancements make such features plausible.

[85] Weather balloons are filled with helium or hydrogen and can reach altitudes of 20-40 km, carrying instruments to measure temperature, humidity, and atmospheric pressure. Payloads often come down far from the launch site due to high-altitude winds.

combination of *activated carbon* and a small UV-C light source to kill bacteria and remove impurities. They tested it on a camping trip with friends.

Leo handed out cups of purified water, giving a mock sales pitch. "Guaranteed to be fresher than a mountain spring."

Mara sipped and smiled. "Not bad. Tastes like victory and minimal gastrointestinal risk."

One friend jokingly dipped a cookie in it, claiming it tasted better too. Newton, ever curious, sniffed a cup but quickly turned away, unimpressed.

Leo chuckled. "Can't win them all."[86]

The Night of the Laser Show Proposal - Leo wanted to propose again—this time in a more flamboyant, light-hearted manner. He set up a laser light show in their backyard, synchronized to project moving constellations that told their story. The last formation was two figures holding hands under a cascade of shooting stars.

As the lights danced, Mara's eyes glistened with laughter. "Leo, you're one photon short of being the brightest star in my life."

He dropped to one knee, holding out a glowing ring. "Mara, let's be a binary star system. Partners in orbit forever."

Newton meowed, unimpressed by the sappy scene, but the couple barely noticed as Mara hugged Leo tightly and whispered, "Yes, again."[87]

---

[86] Activated carbon removes contaminants through adsorption, while UV-C light disrupts the DNA of harmful microorganisms, preventing them from reproducing. This combo is used in many water purification systems today.

[87] Lasers can be used to project intricate patterns and images by directing the beam through a rapidly moving mirror assembly, creating light displays. Their applications range from entertainment to precise industrial cutting and medical procedures.

The Quantum Cat Paradox

Leo's latest obsession was quantum mechanics, inspired by the notorious Schrödinger's cat thought experiment. He couldn't resist joking that Newton might be the feline embodiment of quantum superposition—both a diligent lab assistant and a chaotic menace at the same time.

During one of their brainstorming sessions, Leo proposed building a quantum simulation display that would help explain complex concepts. Using an array of LEDs and a simple computer chip programmed to simulate electron spin states, he created an interactive panel where touching a light would shift its colour to show how particles could be entangled or in superposition.

Mara watched as Newton pawed at the glowing display, triggering lights to blink between blue and red. "He's collapsing the wave function every time he taps," she said, laughing.

Leo grinned. "Or he's just determined to prove we should've named him Schrödinger."[88]

The Case of the Self-Reversing Drone - With drone technology taking off, Leo couldn't resist building his own. He attached cameras, sensors, and even a small AI chip programmed to map its surroundings and correct its flight path. However, he might have been overly ambitious with the AI coding.

---

[88] Schrödinger's cat is a thought experiment illustrating quantum superposition, where a particle can exist in multiple states until observed. It's represented by a hypothetical cat that could be both alive and dead in a sealed box until the box is opened.

During its first test flight, the drone, affectionately named "WingIt," seemed to navigate well until it encountered its reflection in a window. Mistaking it for a threat, it went into defensive mode, doing rapid 180-degree turns and barrel rolls, and zooming straight back to Leo and Mara.

Mara ducked as WingIt hovered in front of them, spinning like it had just finished a coffee-fuelled tango.

"I think it's scared of itself," Mara said, barely containing her laughter.

Leo scratched his head. "Note to self: AI needs to know the difference between threats and mirrors."[89]

The Workshop Jello Incident - One day, Leo decided to host a workshop on *rheology*, the study of how materials flow, specifically non-Newtonian fluids. He mixed up a batch of cornstarch and water to create oobleck, a substance that acts as a solid under force but flows like a liquid when handled gently. To make it more entertaining, he added food colouring and poured it into a tray.

Mara dipped a finger into the colourful mixture, which turned solid when she tried to poke it but dripped through her fingers when she relaxed. "This is both oddly satisfying and infuriating," she said.

Leo attempted to walk across a larger pool of the concoction as part of the demonstration. He managed two confident steps before hesitating, causing his legs to sink knee-deep, resulting in a slapstick-style extraction by Mara.

---

[89] Modern drones use AI and machine learning for tasks such as obstacle avoidance, mapping, and autonomous navigation. They rely on computer vision algorithms to analyse surroundings, which can sometimes lead to funny interpretations of reflective surfaces.

"You've mastered the art of quicksand theatrics," she teased, smirking.[90]

The Jellybean Reactor - Leo's passion for renewable energy inspired him to work on a playful prototype—a "bio-battery" that extracted energy from natural sugars. He gathered a bag of jellybeans and set up electrodes connected to a simple microbial fuel cell.

Mara watched as he plugged in a small LED bulb, which began to glow faintly. "You're telling me we could run a light on candy?"

Leo beamed. "Candy power—sweet, sustainable, and slightly impractical."

Newton, clearly uninterested in scientific breakthroughs that didn't involve lasers or feathers, swatted a jellybean across the table, sending it skittering into the fuel cell and causing a brief flicker of intense light.

"Note: Include Newton-proofing in version two," Mara said, giggling.[91]

The Zero-Gravity Bowling Game - For their next adventure, Leo and Mara managed to join a parabolic flight experience, where passengers experience short bursts of weightlessness. Leo brought along small rubber balls and a portable "space bowling" set to test fluid dynamics and motion under zero-gravity conditions.

---

[90] Non-Newtonian fluids change viscosity based on the force applied. Oobleck is an example of a shear-thickening fluid, becoming more viscous when agitated. This principle is being studied for body armour that remains flexible until impact.

[91] Microbial fuel cells use bacteria to break down sugars and release electrons, generating electricity. While currently not powerful enough to light homes, they showcase how organic waste could become a renewable energy source in the future.

When the plane reached the zero-gravity phase, they released the bowling ball, which floated lazily toward the pins. Mara attempted a perfect throw, only for the ball to hover mid-air and drift sideways, knocking a pin down gently like an apologetic cat paw.

Newton, safely on the ground but watching their live stream, seemed thoroughly unimpressed, lying on Leo's notes like a paperweight.

"Turns out gravity isn't just a buzzkill; it's the bowling champ," Leo chuckled.[92]

The Dual-Purpose Lunchbox - One day, Leo received a challenge from a classmate to build a lunchbox that could keep food warm on one side and cool on the other. Using *thermoelectric materials*, which generate heat or cold based on electric current direction, Leo and Mara got to work.

The final prototype, equipped with a small battery, worked impressively well. One side kept Mara's sandwich warm, while the other kept Leo's fruit cool. However, they hadn't accounted for Newton, who decided the warm side was his new sleeping spot.

When Newton's tail triggered a button, reversing the temperature settings, the cat woke up with a very disgruntled yelp.

Mara laughed. "Looks like Newton prefers room temperature."[93]

---

[92] Parabolic flight simulates zero gravity by flying in a series of arcs. During the top of each arc, passengers experience microgravity for about 20–30 seconds, allowing them to conduct unique physics experiments.

[93] Thermoelectric materials use the Peltier effect, where applying voltage creates a heat difference between two sides of a semiconductor, allowing one side to warm up while the other cools down.

The Molecular Gastronomy Mishap

Leo's newfound fascination with *molecular gastronomy* led to a memorable dinner party for their friends. Armed with liquid nitrogen and agar-agar, he aimed to create edible spheres of tomato soup that would burst in the mouth. Mara watched warily as he dropped small droplets into a chilled bowl.

"Are we sure this isn't just food science masquerading as an experiment?" she joked.

Leo smirked. "Hey, science is the seasoning of life."

As guests gathered and sampled the 'tomato caviar,' one sphere dramatically burst prematurely, splattering soup onto an unfortunate bystander. The room erupted into laughter, and Newton, ever the opportunist, darted in to lap up the remnants.

"It's not a dinner party until someone gets a faceful of soup," Mara quipped, trying to hand the guest a napkin.[94]

The Holographic Plant Assistant - Leo decided to combine his interest in botany with technology by developing a holographic assistant that monitored plant health. Using a miniature projector and sensors for humidity, light, and temperature, he created a small hologram named "HoloGreen" that alerted the user when the plants needed water.

---

[94] Molecular gastronomy uses chemical and physical transformations of ingredients to create novel textures and flavours. The process of spherification involves sodium alginate and calcium chloride, resulting in a thin membrane encapsulating liquid that pops when bitten.

Mara was sceptical until one day HoloGreen projected itself over her laptop, a small watering can icon blinking. "Your basil plant is parched, human," it announced in a surprisingly chipper tone.

Leo laughed. "See, even technology thinks we should be better plant parents."

Newton, intrigued by the sudden burst of holographic light, pounced on HoloGreen, only to pass right through it and land on the keyboard, sending a flurry of random emails.

Mara sighed. "Newton's review: one star, not edible."[95]

The Solar-Powered Marshmallow Roaster - A weekend camping trip with friends led Leo and Mara to test out a portable solar-powered marshmallow roaster. Using concave mirrors and a small solar panel, Leo set up the contraption to focus sunlight and toast marshmallows in minutes.

One of their friends laughed as the marshmallow began to roast perfectly. "You've turned camping into an engineering challenge. What's next, solar-powered bug spray?"

Leo grinned, adjusting the angle. "Give me a day, and we'll talk."

Newton, lounging nearby, observed with mild interest until a faint whiff of caramelizing sugar had him nosing around for treats.

---

[95] Holography relies on the interference of light waves to create a three-dimensional image. While consumer-grade holograms are still developing, applications in augmented reality and plant monitoring with simpler displays are increasingly common.

Mara, marshmallow in hand, raised an eyebrow. "Newton, this is not a self-service buffet."[96]

The Kinetic Art Installation - For a local science fair, Leo decided to build an art installation showcasing kinetic energy principles. He rigged up a series of pendulums, levers, and gears that created a moving mural when set in motion. As the gears turned, different segments flipped, forming a picture of a cat that looked suspiciously like Newton.

Mara noticed the resemblance and laughed. "Let me guess, it's entitled 'Chaos in Motion'?"

Leo nodded. "Inspired by our very own furry overlord."

During the fair, a young visitor accidentally triggered the display's hidden mode, causing the mural to animate with Newton 'chasing' a ball across its panels. The crowd's laughter filled the room, and even Newton himself seemed to approve, eyeing the contraption as if considering how to conquer it.[97]

The Misadventures with Smart Fabrics - Leo's experiments with *conductive fabrics* led to a prototype jacket that could display LED messages controlled by a mobile app. Mara and their friends joked that he'd built the world's first texting coat.

At their next gathering, Leo set the jacket to flash "Hello, World!" and wore it proudly. Mara, not missing an opportunity, changed the message mid-conversation to read, "Ask me about my cat."

---

[96] Solar energy can be harnessed through parabolic mirrors and solar panels to concentrate heat for cooking. The use of reflective surfaces to focus sunlight is employed in solar cookers, which are used in areas without access to traditional cooking fuel.

[97] Kinetic art uses movement as part of its expression, powered by wind, solar energy, or mechanics. The principles of energy transfer and pendulum motion make such displays possible, illustrating conservation of energy and momentum.

Unaware of the prank, Leo found himself surrounded by people inquiring about Newton. When he glanced down and read the new message, he burst into laughter.

"Looks like Newton's got a better PR team than I do," he said, shrugging and accepting his fate as the unofficial spokesperson for his cat.[98]

The Miniature Particle Accelerator - Leo's crowning project for his final research report was an educational miniature particle accelerator model. Using electromagnetic coils and small metal spheres, he demonstrated how particles could be accelerated in a loop and collide to release energy.

At the grand presentation, the tiny spheres zipped around the model, but an unexpected voltage surge caused one to fly off, landing squarely in Newton's food bowl. The cat's wide eyes seemed to say, "Why do things keep landing in my lunch?"

Mara smirked, whispering, "Looks like the universe is conspiring to keep Newton involved."

Leo shrugged, grinning. "A particle accelerator with Newton's seal of approval. That's next-level validation."[99]

The Hilarious Conclusion to Leo's Expedition - Leo and Mara's shared adventures in science continued to blossom alongside their relationship. From holograms that spoke with cheeky personalities to gravity-defying drone dances, each project brought laughter, challenges, and a touch of chaos to their lives. With Newton supervising (or sabotaging) each new

---

[98] Smart textiles incorporate electronic elements such as sensors and LED circuits into fabrics. These textiles are used in wearable technology for everything from monitoring health to making fashion statements.

[99] Particle accelerators use electromagnetic fields to accelerate charged particles to high speeds and direct them to collide. The Large Hadron Collider (LHC) is the most famous example, crucial in discovering the Higgs boson.

endeavour, they learned that science was not just their work but their way of celebrating life.

At their final joint presentation before graduation, they concluded with a slide titled, "Science Is a Cat-Chasing Adventure: Unexpected, Illuminating, and Sometimes Ridiculous." Newton sat on the stage, apparently soaking in the applause as if it were all for him.

Leo turned to Mara with a smile. "Here's to the next experiment."

Mara's eyes sparkled. "May it be as unpredictable as Newton."

And with that, they embarked on their next chapter, a story where science and humour intertwined seamlessly, proving that even the most serious studies could—and should—be met with a grin.

The Great 3D Printer Fiasco

With Leo and Mara now solidly established as co-creators in both research and mischief, they decided to delve into the possibilities of *3D printing*. Their first project was ambitious: designing modular, snap-together structures that could be used to build emergency shelters. The printer, lovingly named "Layer-by-Layer Larry," had a tendency to glitch, adding an unexpected layer of suspense to their experiments.

On one memorable night, Leo programmed Larry to print a scale model of a geodesic dome. The machine whirred confidently for the first few hours, but at some point in the middle of the night, it hiccupped and started printing in mid-air. By morning, an odd, spaghetti-like web covered the lab like a bizarre modern art installation.

Mara arrived, coffee in hand, and blinked at the scene. "Larry's gone rogue. This is how the robot uprising begins, isn't it?"

Leo laughed as he untangled himself from a filament loop. "We'll be remembered as the first victims of abstract art warfare."[100]

Newton's Magnetism Mishap - Leo and Mara were testing the properties of neodymium magnets for a new wireless power transfer project. Newton, ever curious, batted one of the small but powerful magnets onto the floor. As they bent down to pick it up, another magnet from the table jumped to join its

---

[100] 3D printing, or additive manufacturing, creates three-dimensional objects by laying down successive layers of material. It's used for everything from prosthetics and aerospace components to creative art installations. But calibration is key; without it, printers can misalign, leading to mid-air printing mishaps.

friend, trapping a stray paper clip and Leo's notes between them in a bizarre sandwich.

"Newton, are you secretly practicing magnetic levitation?" Mara asked as the cat stared with wide, guilty eyes.

Leo tried to separate the magnets. "These things are stronger than my will to resist snacks at 3 a.m.!"[101]

The Polymorphic Polymer Party - For another project, Leo and Mara experimented with *shape-memory polymers*— materials that could return to their original shape after deformation when exposed to certain stimuli, like heat or light. The duo created a small, fun prototype: a flower that would bloom when exposed to a heat lamp.

As they presented it at a science fair, Leo proudly said, "With just a bit of heat, nature's miracle can be replicated. Watch this!" He turned on the lamp, and the flower unfurled gracefully.

Until it didn't. A minor programming error made the petals snap back a little too forcefully, launching a small rubber band directly into the audience. It landed with pinpoint precision in Newton's fur, who yowled dramatically and darted offstage, dragging the band with him.

Mara stifled laughter. "Nature can be aggressive too, I guess."[102]

---

[101] Neodymium magnets are rare-earth magnets known for their impressive strength. Despite their small size, their magnetic pull is much stronger than typical ferrite magnets and can be used in electric motors, magnetic fasteners, and wireless power devices.

[102] Shape-memory polymers respond to environmental changes, making them ideal for medical stents, self-healing materials, and deployable structures in aerospace. They work by returning to a "programmed" shape when triggered by temperature or other stimuli.

The Accidental Biofuel Discovery - Leo's passion for sustainability led him to experiment with biofuels made from kitchen waste. After an extensive evening of grinding leftover vegetable peels and mixing them with enzymes, they managed to create a small, workable fuel cell.

Mara watched as Leo lit a tiny flame from their makeshift burner. "We're one step closer to eco-friendly road trips!"

Before they could celebrate further, Newton jumped onto the bench, tipping the container and sending a cascade of liquid onto the floor. Luckily, Leo had a towel ready, but Mara couldn't help but quip, "Newton's just making sure we remember: don't cry over spilled biofuel."[103]

The Holographic Newton Display - In their ongoing bid to integrate their love for science with everyday life, Leo created a holographic projector that displayed Newton's image—an endless loop of him batting at floating butterflies. The idea was to test projection angles and light reflections on non-standard surfaces.

Newton, upon seeing his holographic twin, reacted predictably: by charging directly at it. The cat skidded through the hologram and landed in a pile of papers, scattering Leo's research on polymer elasticity across the room.

Mara raised an eyebrow, picking up a sheet. "Looks like Newton's testing out Newtonian mechanics on your research notes."

---

[103] Biofuels are produced through biological processes, like the fermentation of plant materials. They're renewable and can reduce greenhouse gas emissions compared to traditional fossil fuels. Second-generation biofuels, made from non-food biomass, are considered more sustainable.

Leo smirked, scooping up their feline tornado. "If we could patent his chaos as renewable energy, we'd be set for life."[104]

The Conclusion to Leo's Lab Year - The year ended with a grand showcase where Leo and Mara presented a compilation of their projects: from Newton's magnetic adventure to their unexpectedly humorous 3D printer failures. The holographic cat display was a crowd favourite, and their biofuel project drew interest from several professors for further development.

As the applause echoed and friends gathered around them, Mara nudged Leo. "We made it through another year of scientific mayhem."

Leo smiled, eyes twinkling. "And Newton made it through another year of being our unplanned, furry collaborator."

From holograms to polymers, Leo's journey in the realm of science was a testament to resilience, curiosity, and, most importantly, the laughter that came from the joy of discovery. And with Mara by his side and Newton always a step away from the next mishap, the adventure was bound to continue.

---

[104] Holographic projection works by manipulating light to create three-dimensional images that appear to float in space. Although still in its experimental phase for broader uses, advancements are making holography a staple in education, art, and medicine.

The Levitation Experiment Gone Wild

Leo's fascination with magnetism evolved into an attempt to create a *levitating platform*. The principle was straightforward: use repelling magnetic forces to float a small disk in mid-air. With precise placement and powerful neodymium magnets, Leo and Mara were close to achieving the perfect balance.

"Ready to see Newton's new hovercraft?" Leo joked, adjusting the platform.

Mara crossed her arms, smirking. "I'm ready for Newton to finally believe he's capable of flight."

After a few calibrations, the platform wobbled and lifted a couple of inches. Newton, intrigued by the hovering device, circled it like a lion stalking its prey. Before anyone could react, the cat jumped onto the platform. The sudden added weight made the platform spin rapidly, causing Newton to experience what could only be described as a brief zero-gravity amusement park ride.

"Newton's testing the centrifugal force, clearly," Mara said, holding back laughter as the cat jumped off, fur on end, staring at the platform in betrayed disbelief.[105]

The Accidental Battery-Baking Incident - Leo's research in renewable energy led him to experiment with *solid-state batteries* using electrolyte gels. The duo worked meticulously, mixing components that could potentially enhance battery

---

[105] Magnetic levitation relies on the repelling or attracting forces between magnets. Maglev trains use this principle to reduce friction, allowing for smooth, high-speed travel. The interplay of magnetic fields and stability—using diamagnetic materials or precise feedback mechanisms—keeps levitating objects steady.

efficiency and safety compared to conventional lithium-ion batteries.

One evening, Leo popped the gel into an oven for controlled heating, which was supposed to help the substance set. However, he'd forgotten to adjust the temperature, and a faint smell of something between burnt toast and a chemistry lab gone wrong filled the room.

Mara entered, sniffing the air. "Why does it smell like we're baking batteries instead of cookies?"

Leo opened the oven, coughing. "Because we are. But on the bright side, we now know the ignition point of electrolyte gels."

Newton, nose twitching, peeked into the kitchen, then promptly left with an expression that screamed, *even I know better.*[106]

The Water-Powered Speaker - In a bid to combine sustainability with entertainment, Leo worked on a *hydro-powered speaker* system. He rigged a miniature turbine that generated electricity as water flowed through it. The idea was to play music while demonstrating hydropower in action—a perfect synthesis of fun and science.

The first test was a raucous success, with the turbine spinning wildly as the water gushed and Beethoven's "Ode to Joy" rang out. That is, until Newton, curiously swatting at the water, sent the carefully directed stream veering into Leo's face.

---

[106] Solid-state batteries use solid electrolytes, making them less flammable and potentially more energy-dense than their liquid counterparts. They are a significant focus for research in electric vehicles and other energy storage solutions.

Mara burst out laughing, barely able to speak. "Hydropower: one. Leo: zero."

Leo wiped his face, grinning. "I'm making a note: 'cat-proofing' needs to be added to the design phase."[107]

The DIY Mars Rover - Inspired by news of the Perseverance rover and its remarkable discoveries on Mars, Leo set out to create a mini-version with similar autonomous exploration capabilities. It had sensors for detecting terrain and even a small claw for collecting samples—mostly small rocks or, as Newton discovered, cat toys.

During a live demonstration for their classmates, the rover wandered off course and started following Newton, who quickly adapted to having an automated admirer. The cat led the rover on an improvised parade across the room, with it chirping mechanical beeps of acknowledgment whenever Newton swatted at it.

Mara couldn't help but narrate. "And here we see the Rover-Newton symbiosis: a rare collaboration in the wild."[108]

The Home Chemistry Stand-Up Show - Leo and Mara decided to host a "Science Comedy Night," where they shared humorous anecdotes and live experiments with their classmates. One standout moment involved demonstrating how *sodium bicarbonate (baking soda)* reacts with *vinegar* to create carbon dioxide gas and inflate a balloon.

---

[107] Hydropower involves converting the kinetic energy of flowing or falling water into mechanical energy and then into electrical energy. Small-scale hydro systems, like water wheels and micro-hydro turbines, are often used for local power generation and educational demonstrations.

[108] Mars rovers like *Perseverance* and *Curiosity* are equipped with cameras, spectrometers, and robotic arms for geological sampling. They employ advanced AI to navigate autonomously, avoid hazards, and make decisions on which samples to collect.

Leo set up the demo with extra flair. "Ladies and gentlemen, I present to you the world's most dramatic balloon inflation. Nature's proof that even the simplest things can blow up out of proportion."

He poured the vinegar into the flask, and the balloon inflated rapidly. Unfortunately, Mara had tied it a bit too tight, causing it to pop and shower the front row with droplets. The startled laughter that followed only added to the chaotic charm.

Newton, watching from a safe distance, merely flicked his tail, unimpressed by such predictable outcomes.[109]

The Aftermath: Winning the Unintended Awards - Leo's collection of experiments and their mishaps turned into an accidental sensation. The school newspaper dubbed him "The Unintentional Entertainer," while professors lauded his innovative thinking and ability to engage people with science through humour. Even Newton was awarded an honorary "Paws of Participation" certificate.

Leo accepted his reputation with grace. "If the world remembers me as the scientist who made people laugh and think, I'm happy with that."

Mara raised a toast at their celebratory get-together. "To science that sparks curiosity, laughter, and the occasional flying marshmallow."

---

[109] The reaction between sodium bicarbonate and acetic acid (in vinegar) releases carbon dioxide gas. This experiment is a simple demonstration of an acid-base reaction and the production of gas, illustrating basic principles of chemistry and pressure.

The Solar-Powered S'mores Maker

Leo's latest inspiration came from his desire to blend leisure and sustainability. He designed a *solar-powered oven* capable of reaching temperatures sufficient to cook basic foods. The plan? Make s'mores and share them during a campus picnic.

Mara helped him construct the setup, which included reflective panels to concentrate sunlight onto a small insulated box. As they positioned it on the university lawn, Newton watched from a nearby patch of shade, eyes half-closed but ears twitching, as if preparing for another impromptu science show.

"Alright, ready for some guilt-free, eco-friendly desserts?" Leo announced as he placed a marshmallow and chocolate sandwich inside the contraption.

For a while, everything went smoothly. The marshmallow puffed up like a fluffy cloud, the chocolate softened to gooey perfection, and Leo prepared to serve the first solar-s'more. Just as he turned to Mara to celebrate, a sudden cloud passed over, cutting off the sunlight.

"Perfect timing," Mara said, fighting back a laugh as the once-promising s'more cooled and began hardening into an unappetizing block.

Newton chose that moment to swat at a stray graham cracker, sending it skittering across the grass. The cat then followed with regal indifference, claiming it as his new toy.

Leo sighed, eyes crinkling with humour. "Note to self: add a backup heating element for cloudy days."[110]

The Talking Plant Experiment - Leo's next project involved testing theories on *plant communication* and responses to stimuli. Inspired by studies that suggested plants could react to sound waves, he decided to build an experiment where plants "listened" to different types of music.

Mara, always game for creative research, curated a playlist that ranged from classical music to heavy metal. Leo set up the speakers and positioned a row of plants, each labelled with genres like "Beethoven's Basil" and "Metallica's Mint."

After a week of exposure, the results were mixed. The classical-music plants seemed to grow slightly taller, while the heavy metal plants developed a peculiar lean, as if headbanging had become a part of their growth pattern. Newton, unimpressed by the array of musical vegetables, sprawled out in front of the jazz section, apparently preferring smooth saxophone solos.

One night, during a routine check, Leo joked, "If these plants start requesting encore performances, we'll know we're onto something."

"Or if Newton starts playing guitar," Mara added with a grin.[111]

The Liquid Nitrogen Ice Cream Night - Leo and Mara decided to host an ice cream night for their friends, but with a twist:

---

[110] Solar ovens use reflective surfaces to direct sunlight into an enclosed space, raising the internal temperature high enough to cook food. The principle relies on the greenhouse effect—sunlight enters, is absorbed, and then converted to heat, which is trapped inside.

[111] Research on how plants respond to sound has shown that they can react to certain frequencies, potentially impacting growth and development. Though the science isn't fully settled, the idea of plants "listening" to their environment suggests they're more dynamic than once thought.

using *liquid nitrogen* to freeze the ingredients. They prepped a large mixing bowl, poured in the ice cream base, and, donning protective goggles, began adding the liquid nitrogen.

A dramatic fog rolled out as the liquid nitrogen boiled off, drawing excited "oohs" and "aahs" from their guests. Newton, on high alert at the strange sound, peered cautiously from under the table, eyes wide with suspicion.

"Safety tip: never try this at home without supervision," Leo said, stirring the rapidly cooling mixture. The result? Smooth, creamy ice cream in record time.

Mara handed out bowls, adding, "It's like sorcery, but science-approved."

Suddenly, one friend leaned too close to the bowl and let out a startled yelp when the nitrogen fog curled around their face. Leo chuckled. "See? Science that takes your breath away—literally."[112]

The Hydrogen Balloon Hoax - In a particularly mischievous mood, Leo decided to pull a light-hearted prank involving *hydrogen-filled balloons*. He made sure they were safely handled and far away from any actual ignition sources, but the concept was simple: fill balloons with hydrogen, attach notes like "This balloon is lighter than your last physics exam," and set them loose around campus.

Mara walked into the lab and saw the notes. "Should we expect a series of calls from the campus safety office, or...?"

Leo grinned. "They'll learn that it's just basic science. Hydrogen is lighter than air and loves to rise."

---

[112] Liquid nitrogen is at a temperature of around -196°C (-321°F) and is used in culinary settings to rapidly freeze food, creating a smoother texture. The liquid evaporates quickly when exposed to room temperature, producing dense, chilly fog.

Later that day, confused yet entertained students watched as the balloons floated up and gently drifted away, sparking conversations about buoyancy and elemental gases. The chemistry professor, passing by, chuckled knowingly. "Leo strikes again," he murmured.

Newton watched the scene unfold from a window ledge, tail flicking as if contemplating how to reach one of those strange floating toys.[113]

The Invisible Ink Reveal - For one of their research papers, Leo and Mara used *invisible ink* to add hidden jokes and footnotes. They prepared it using a simple mixture of lemon juice and water, which turns visible when heated. During a presentation, they revealed the secret, passing a hairdryer over sections of their paper to expose hidden comments like, "Newton's legal team did not approve this hypothesis."

The class erupted in laughter as more phrases appeared, each one funnier and more absurd than the last.

Afterward, one of their professors approached, smiling. "Creative presentation. You might be the first students to submit a paper that's also a comedy set."[114]

The Grand Solar Showdown - At the end of their final term before summer, Leo organized a small event he dubbed "The Solar Showdown," inviting classmates to create solar-powered contraptions and race them. Mara and Leo contributed a solar-powered Newton-mobile—essentially a tiny cart designed for

---

[113] Hydrogen gas is much lighter than air and can make balloons float similarly to helium. However, hydrogen is flammable and is usually avoided in casual balloon use, with helium being the safer choice for party balloons.

[114] Invisible ink made from lemon juice works because it weakens the paper fibres where it's applied. When heat is added, the lemon juice oxidizes faster than the surrounding paper, making it appear brown.

the cat, equipped with a small canopy and cushions for comfort.

During the race, Newton, perched regally, gazed at the other makeshift solar vehicles, which sputtered or wobbled as clouds played hide-and-seek with the sun. When the sun finally beamed through, Newton's cart accelerated with a lurch, causing him to grip the sides with wide eyes before jumping off and letting it roll to victory.

Leo threw his arms up triumphantly. "We've officially created the first cat-approved solar transport!"

Mara's laughter echoed across the field. "Let's hope the world is ready."[115]

---

[115] Solar panels work by converting sunlight into electricity using the photovoltaic effect. When photons strike the surface of a solar cell, they displace electrons, creating an electric current that can be harnessed to power devices or charge batteries.

The Coffee Catalyst Experiment

As Leo, Mara, and their research circle entered the final stretch of their university years, coffee became a necessity and a running joke. One day, Leo proposed a novel idea: develop an experiment on the chemical reactions of caffeine and its potential as a catalyst in simple chemical processes. It was, as Mara dubbed it, "the most college-level science project yet."

Leo's hypothesis was straightforward: caffeine, as a stimulant, could be used to promote certain reactions. He set up an experiment involving hydrogen peroxide and potassium iodide, with a few drops of highly concentrated caffeine solution added for flair. As expected, the mixture bubbled more vigorously than usual.

"Is that...?" Mara squinted at the frothy concoction.

Leo nodded. "The caffeine isn't reacting as a true catalyst would, but it's increasing the overall reaction rate a bit. Or it's placebo science. Who's to say?"

Newton sniffed the bubbling container and made a quick retreat. It seemed even he knew not to trust caffeine experiments.[116]

The Mechanical Pencil Arm-Wrestling Challenge - One afternoon, Leo had the brilliant idea of building a mechanical arm powered by a simple pneumatic system. His goal? Create an arm that could engage in an *arm-wrestling match* with a

---

[116] Caffeine acts as a stimulant by blocking adenosine, a neurotransmitter that makes you feel sleepy. While it doesn't function as a chemical catalyst in typical reactions, it can influence the metabolism and reactions within biological systems.

human. It was a feat of materials science and robotics that combined precision-engineered joints, springs, and pressure chambers.

"Are you telling me you built this just to see if you could beat your classmates in arm-wrestling without lifting a finger?" Mara asked, eyebrow raised.

"Precisely. And because it's cool," Leo replied with a grin.

When it came time to test it out, Leo faced off against the machine in front of their friends. With a dramatic hiss, the pneumatic arm clamped down, slowly but surely winning the match. Everyone burst into laughter, some jokingly asking if Leo would now train to defeat his own creation.

Newton, unimpressed, attempted to play with the arm's mechanical fingers, adding a layer of unintended humour as the robotic hand tried to keep up with the curious cat.[117]

The Campus-Wide Scavenger Hunt for Rare Metals - Leo's fascination with materials science led him to orchestrate a campus scavenger hunt themed around discovering "rare metals." He hid small, labelled pieces of replica metals—like tungsten, titanium, and cobalt—around the campus. Each clue involved a scientific riddle that referenced the unique properties of these metals.

"The next clue might be 'heavy,' but don't be fooled—it's stronger than steel," Mara read aloud from her clue sheet.

One group found their clue attached to the gym's dumbbells (a nod to tungsten's high density), while another discovered

---

[117] Pneumatic systems use compressed air to generate mechanical motion. They are often used in robotics for lightweight, quick actions, unlike hydraulic systems, which use liquids and provide greater force but are slower.

titanium hidden in the bike racks. The highlight was Newton batting around the "gold" clue, carrying it like a prized catch.

Leo announced the prize: a miniature periodic table with real samples of some metals embedded. Everyone agreed it was the nerdiest, most engaging scavenger hunt they'd ever participated in.[118]

The Surprise Robo-Band Performance - Inspired by their experiments in robotics, Leo and Mara decided to create a small band of robot musicians for an end-of-term celebration. They programmed a drum machine, a mechanical arm that strummed a guitar, and even an automated maraca shaker. Newton was given a role too: pressing a big, red button that started the performance.

When the night arrived, the makeshift robot band was set up under a canopy of string lights. Leo introduced the act with dramatic flair. "Ladies, gentlemen, and Newton, presenting: 'The Circuit Shakers'!"

Mara added, "No cover fee, but tips in the form of positive feedback are welcome."

As the first notes played—robotically perfect yet charmingly awkward—the crowd cheered. Halfway through, Newton, true to his mischievous nature, swatted the button twice, triggering a drum solo that had the whole crowd in stitches.[119]

The Great Chemistry Showdown - As a final act of scientific theatre, Leo proposed a chemistry demonstration event where

---

[118] Tungsten has the highest melting point of all metals and is often used in light bulb filaments. Titanium is known for its high strength-to-density ratio and is used in aircraft and medical implants due to its lightweight nature and biocompatibility.

[119] Robotic bands are a real concept used to explore AI in music. Robots can be programmed to play instruments using actuators, sensors, and pre-defined musical algorithms. While lacking human nuance, they showcase the precision of automated systems.

participants showcased flashy chemical reactions. His presentation? A modified *elephant toothpaste* reaction with colourful dyes added for effect.

Leo set up on stage with Mara acting as his "safety manager," complete with oversized goggles and a spray bottle filled with water for comedic effect. The crowd watched with anticipation as Leo mixed the hydrogen peroxide, dish soap, and potassium iodide, each coloured solution blending into a rainbow effect. The reaction began slow, then surged into a vivid cascade of foam.

"Let's just hope it doesn't reach Newton's perch," Mara whispered.

Newton, from his spot on the windowsill, observed with narrowed eyes, ready to leap if needed.

As the foam continued to spill out in towering waves, Leo declared, "It's not just an experiment; it's a cleaning service!"[120]

The Midnight Laser Maz - For their last hurrah before the final exams, Leo and Mara turned one of the campus labs into a maze using *laser pointers and mirrors*. The objective was to navigate the room without breaking any of the laser beams, Mission Impossible-style.

Students queued up, laughing nervously as they twisted, ducked, and contorted their way through. Leo and Mara monitored the maze from a control panel, occasionally shifting a mirror slightly for an extra challenge. Newton, naturally intrigued by moving lights, decided to join in,

---

[120] The "elephant toothpaste" reaction is a rapid decomposition of hydrogen peroxide using potassium iodide as a catalyst, releasing oxygen gas that gets trapped in soap bubbles. This reaction demonstrates the breakdown of hydrogen peroxide into water and oxygen, a principle often used to teach catalytic reactions and gas production.

creating chaos as he batted at the lasers and set off every sensor.

Mara threw up her hands. "Technically, Newton wins the night with the most laser trips."

Leo announced, "And he remains undefeated in the 'Cat vs. Science' league."[121]

The Final Goodbye Party - Their time at university culminated in a grand farewell party, where friends shared stories of Leo's many escapades—from exploding marshmallows to Newton's stardom. Professors joined in, applauding Leo and Mara's contributions to the campus's spirit of inquiry and fun.

Newton, wearing a tiny graduation cap, received a standing ovation for his unintended but essential role in their journey.

Leo raised a glass. "To every hypothesis tested, every lab table conquered, and every time Newton taught us that curiosity isn't just for humans."

The crowd cheered as Mara whispered, "So, what's next?"

Leo smiled, eyes sparkling with excitement. "Everything, Mara. Absolutely everything."

---

[121] Laser pointers emit coherent light that can be reflected and directed using mirrors. In labs, lasers are used for precise measurements, spectroscopy, and even surgeries, showcasing the incredible versatility of light.

The Grand Summer of Experiments

With university life wrapped up and a full summer stretching before them, Leo and Mara found themselves itching for more scientific escapades. Leo's curiosity remained boundless, and Newton's antics were, as ever, the unexpected twist in their narrative. This time, however, Leo wanted to push his boundaries beyond traditional projects. The summer would be an *experiment of life*, packed with new discoveries and an abundance of humour.

Project 1: The Quantum Ice Experiment - Leo's latest fascination was with the peculiar properties of *supercooled water*, which stays liquid even below its freezing point until it is disturbed. He proposed a simple yet captivating demonstration at the local science fair.

"We'll take bottles of water, chill them below zero without freezing, and then, with a tap, turn them instantly into ice. We'll call it 'The Quantum Chill,'" Leo declared to Mara.

On fair day, Leo set up a row of pre-chilled water bottles. As the crowd gathered, he gave a dramatic countdown, tapped the first bottle, and—nothing happened. Silence fell over the audience.

Mara leaned in, whispering, "Did we forget the magic words?"

Newton, sensing the tension, jumped up onto the table, knocking over a bottle. To everyone's amazement, the tipped bottle crystallized instantly as it hit the table, eliciting gasps and cheers from the crowd.

Leo grinned. "Note to self: always keep Newton on standby as a scientific assistant."[122]

Project 2: The Non-Newtonian Dance Floor - Inspired by the behaviour of *non-Newtonian fluids*, Leo decided to turn his backyard into an interactive science experiment. He and Mara mixed a large batch of cornstarch and water to create oobleck, a substance that behaves like a liquid under slow movement but solidifies when force is applied.

"We'll make it a dance floor!" Mara suggested, eyes sparkling with excitement. The idea was simple: play music and invite friends to dance on the oobleck. If they moved fast enough, they'd stay on the surface. If they stopped or moved too slowly, they'd sink.

Friends arrived, intrigued and slightly wary. The first beats of an upbeat song echoed through the yard, and everyone joined in, hopping and sliding on the strange, semi-solid surface. Newton watched from a safe distance, pawing at a thin splatter that reached the grass.

As soon as the tempo changed to a slow ballad, participants who failed to adjust quickly enough found themselves knee-deep in the oobleck, laughing and scrambling to get back to solid ground.

Leo laughed so hard he nearly sank himself. "This proves two things: science can be fun, and no one survives slow dancing on oobleck."[123]

---

[122] Supercooled water is liquid water cooled below its normal freezing point. It can remain in this state because the molecules don't have an initial nucleation point to start forming ice crystals. Disturbing the container provides this starting point, causing the water to freeze suddenly.

[123] Oobleck is a non-Newtonian fluid, meaning its viscosity changes under stress. When moving slowly, it acts like a liquid, but when force is applied, it behaves like a solid due to the rearrangement of cornstarch particles.

The Floating Garden Bet - Mara challenged Leo to design a *floating garden* using principles of buoyancy and sustainable water management. Leo loved the idea, but he wanted to take it further and make it humorous. He added miniature sailboats rigged with tiny, wind-powered flags that read "Future Scientists Rule."

Construction involved building a lightweight base from recycled plastic and layering it with soil and hydroponic elements. They populated the floating platform with basil, mint, and a single sunflower, aptly named "Newton's Sun."

At the launch event, a breeze picked up, causing the sailboats to spin wildly, one even flipping over. Newton, perched at the edge of the pond, swatted at a nearby flag as if he'd just discovered his new summer hobby.

"You know, if this gets any more chaotic, we might get a research paper on how not to build floating gardens," Mara joked.

Leo chuckled. "At least Newton's Sun seems unbothered. A true symbol of perseverance."[124]

The DIY Space Balloon Launch - Leo had always wanted to send something into the stratosphere, so when a friend mentioned amateur *weather balloon* launches, he jumped at the chance. The goal was to attach a small camera and a note that read, "Hello, stratosphere! Courtesy of Leo and Mara." Newton's contribution? A cat toy tied to the rig for a touch of humour.

---

[124] Floating gardens, or chinampas, have been used since the time of the Aztecs and work on the principle of buoyancy. The plants absorb water from below while floating structures are built using materials less dense than water.

"Are we sure Newton's okay with us sending a toy into space?" Mara teased as they attached the final elements.

Leo smiled. "He'll get over it when it comes back with cosmic dust."

On launch day, a crowd of friends gathered to watch the balloon, filled with helium, lift the small payload. It soared upward, drifting over trees and out of sight. Hours later, the GPS signal pinged back to Earth, and they retrieved the camera with footage showing the blackness of near space and a tiny, wagging cat toy bobbing in the view.

The video quickly went viral among their circle, complete with captions like, "Newton's first toy to touch the stars."[125]

The Solar-Powered Smoothie Stand - Taking their sustainability theme further, Leo and Mara set up a solar-powered smoothie stand at a local festival. They rigged solar panels to run a small blender, serving smoothies with names like "Photon Punch" and "Electron Elixir."

The first day was smooth sailing, and their booth drew curious onlookers. But halfway through, a cloud cover rolled in, slowing down their operations. Mara raised an eyebrow. "Backup generator?"

Leo smirked. "Nope. Plan B is arm-powered blending."

Friends took turns cranking an old manual blender while Leo narrated. "This is what happens when the sun goes on lunch break."

---

[125] Weather balloons can reach altitudes up to 30 kilometres (about 100,000 feet), where the Earth's curvature becomes visible and the sky appears black due to the thin atmosphere. Helium is used as it is less dense than air and non-flammable, making it safe for such experiments.

Newton lounged under the table, only twitching his tail when an accidental berry dropped near him.[126]

The Unintentional Quantum Cat Toy - Their next invention was an automated laser pointer on a rotating arm for Newton. Programmed with random intervals, it was supposed to be entertaining for the cat and scientifically intriguing to Leo, who wanted to see if randomness could mimic the concept of quantum superposition.

The first test was chaos. Newton, eyes wide and tail fluffed, darted back and forth, trying to catch the elusive red dot that seemed to teleport unpredictably. The room soon echoed with laughter as Newton made exaggerated leaps, narrowly missing shelves.

"Quantum cat's a reality, folks!" Mara announced between giggles.

Newton, realizing the laser was unbeatable, gave it one disdainful glance and left the room, tail swishing with the haughtiness of a creature who knew he was better than quantum tricks.[127]

The summer proved to be everything Leo had hoped for: full of scientific misadventures, breakthroughs, laughter, and the companionship of friends and Newton. As the sun set on their last day before Mara left for her internship in Germany, Leo reflected on the journey.

---

[126] Solar panels convert sunlight into electrical energy using the photovoltaic effect. When photons hit the solar cells, they excite electrons, creating an electric current. Solar energy efficiency varies based on the panel's quality and sun exposure.

[127] Quantum superposition refers to a quantum system's ability to exist in multiple states simultaneously until it's measured or observed. Leo's random laser pattern, though not quantum, humorously nodded to the unpredictability of such phenomena.

"Do you think we've done enough this summer?" Mara asked, glancing at their cluttered workspace filled with half-finished contraptions and note-stuffed boards.

Leo leaned back, smirking. "Enough to fill an entire chapter in 'The Chronicles of Science Misfits.'"

Newton, as if on cue, knocked over a small vial, sending a tiny puff of smoke into the air. They burst out laughing, and Leo added, "And that, my dear Mara, is exactly how we know we did."

The future shimmered with endless possibilities, and Leo was ready for the next phase with a grin, a cat, and his ever-insatiable curiosity.

A Journey Culminating in Stardust

As the summer of experimentation wound down, Leo felt a familiar excitement tinged with the gentle hum of anticipation. His career was advancing rapidly. The whimsical days of impromptu experiments and slapstick science with Newton were woven into a larger tapestry of serious discovery and renowned recognition.

Leo's groundbreaking work in materials science, inspired by everything from Newton's escapades to his moonshot projects with Mara, was being talked about in scientific circles across the globe. His research had taken the properties of *metamaterials* — synthetic composites with unconventional properties like negative refractive indices — and used them to pioneer a method for creating ultralight yet incredibly strong structures. His approach had applications from aerospace to sustainable architecture.[128]

A Proposal Among the Stars - Leo and Mara's romance had blossomed amidst lectures, expeditions, and the thrum of endless experiments. They'd shared a bond that merged curiosity with humour, where conversations seamlessly moved from quantum theories to whether Newton would one day achieve feline omnipotence. One evening, under the clear, star-laden skies of the Swiss Alps — their current destination for a joint materials science conference — Leo felt the moment was right.

---

[128] Metamaterials are materials engineered to have properties not found in naturally occurring substances. They can manipulate electromagnetic waves and have potential applications in creating invisibility cloaks, improved antennas, and lenses with super-resolution.

After a day of presenting their respective research, Leo led Mara to a secluded hill overlooking the mountains. The night air was crisp, and the stars seemed to wink conspiratorially. Newton, who had accompanied them and spent most of the conference charming attendees, snoozed under a nearby bench.

"I think we've discovered almost everything under the sun," Leo said, slipping into a playful tone. "But there's one experiment I'd like to propose that I think we should try."

Mara raised an eyebrow, smirking. "Let me guess, does it involve non-Newtonian fluids and dancing again?"

"Nope, but it involves the two of us," Leo replied, pulling out a small ring that glimmered under the starlight.

Mara's eyes widened, and her laughter turned into tears. "Is this what you've been building with all that hidden lab time?"

Leo nodded. "I present to you a ring made with our carbon synthesis method. Lightweight, hyper-durable, and quite possibly the only ring in the world reinforced with love and a touch of Newton's fur."

Mara laughed and nodded. "Yes! But only if Newton approves."

Newton, as if on cue, opened one eye, let out a disinterested purr, and promptly fell back to sleep.

The Nobel Project - A year later, Leo and Mara found themselves knee-deep in their most ambitious project yet: a sustainable, self-healing composite material capable of repairing microscopic cracks using embedded nanobots activated by light. The applications spanned fields from infrastructure repair to spacecraft skin that could withstand micrometeorite impacts and self-heal in space.

Their shared workspace was a hurricane of notes, gadgets, and odd contraptions. Newton had taken to carrying tiny bolts in his mouth like a foreman supervising construction.

"Why do I feel like we're one malfunction away from accidentally inventing sentient slime?" Mara joked one day, adjusting her lab goggles.

Leo chuckled. "Because we probably are. And when that happens, Newton will have to negotiate a treaty."[129]

Their innovation took the world by storm. The material demonstrated unprecedented resilience during tests: an aircraft wing patched itself mid-experiment while a group of stunned scientists watched. News outlets dubbed it *the beginning of the self-repairing world.*

The Awards Ceremony and the Cat That Stole the Show - A year after the successful implementation of their discovery, Leo and Mara found themselves dressed to the nines at the Nobel Prize awards ceremony in Stockholm. Leo stood backstage, palms sweaty, eyes scanning the crowd for Mara. She caught his eye and mouthed, "You've got this," before gesturing to Newton, who, against all odds, had made his way onto the esteemed guest list wearing a tiny bow tie.

As their names were called, the hall erupted in applause. Leo took Mara's hand, walking to the podium together. He cleared his throat, glancing down at Newton, who had positioned himself proudly at the front of the stage.

---

[129] Nanotechnology in materials science enables the creation of materials with unique properties. Nanobots embedded in substances can be programmed to perform specific tasks, like repairing structural damage, activated by certain stimuli, such as light or temperature changes.

"Firstly, we'd like to thank our most peculiar co-researcher, Newton, whose antics kept our lab lively — and our coffee spilled."

The crowd chuckled, and Newton let out a theatrical meow.

Leo continued, "This journey began as a child's dream of understanding the invisible world of atoms, expanded with countless late-night experiments, and bloomed in a partnership forged in curiosity and a love for discovery. This recognition is not just for us, but for all who wonder, fail spectacularly, laugh, and try again."

They were awarded the prize in materials science, their project lauded as a pioneering leap toward sustainable, intelligent structures.[130]

A Life Full of Laughter, Love, and Science - Post-award life was a balance of groundbreaking research, public speaking engagements, and quiet evenings spent in their garden, now outfitted with solar-powered lights and tiny cat-sized experiments. Leo would often narrate science stories to Mara, with Newton lounging regally as if absorbing the knowledge himself.

As the sun set on their home filled with laughter and the low hum of ongoing experiments, Leo reflected on the journey from a curious kid who played with kitchen magnets to a Nobel laureate sharing a life of discovery and joy.

Mara nudged him. "Penny for your thoughts?"

---

[130] Self-healing materials are at the forefront of innovative research. They incorporate microcapsules or nanobots that release healing agents or reconfigure damaged areas when triggered, inspired by biological systems that repair themselves.

Leo smiled. "Just wondering what Newton's Nobel speech would be if he had one."

Mara laughed. "Something along the lines of, 'Thanks for the treats, but I demand a bigger lab bed.'"

Newton, seemingly understanding, yawned and nestled deeper into his favourite cushion — a contented co-researcher in a home where science and laughter were eternally intertwined.

Epilogue: Newton's Contribution - In an amusing twist, Newton the cat had become somewhat of a mascot for the science community. His viral antics in the lab had inspired a series of children's books called *Newton's Scientific Tails*, teaching kids basic principles of physics, materials, and robotics.

The Nobel Surprise and the Next Chapter

It was a bright, slightly chaotic morning when Leo received the call that would change everything. He was in the middle of adjusting the settings on a laser experiment when his phone rang. It was an unknown number, and Leo, assuming it was another telemarketer hawking laboratory supplies or "the world's best lightbulb," ignored it. When it rang a second time, Mara glanced up from across the room, eyebrow raised.

"Leo, just answer it. It could be important."

With an exaggerated sigh, Leo put the phone on speaker. "Hello, this is Leo."

A crisp, accented voice responded, "Good afternoon, Mr. Leo Magnus. This is the Nobel Committee calling from Stockholm."

Leo nearly dropped the phone, his jaw going slack. Mara's eyes widened, and Newton, who had been playfully batting a laser pointer's beam, froze mid-swat as if sensing the gravity of the moment.

Leo laughed nervously. "Oh, good one! Did Tom from the lab put you up to this? Let me guess, he's recording me for the next 'Gotcha Scientists' viral video."

The voice on the other end didn't waver. "No, Mr. Magnus. This is very real. We are pleased to inform you and Dr. Mara that you have jointly won the Nobel Prize in Chemistry for your pioneering work on self-healing, nanobot-embedded materials."

Mara gasped, clutching Leo's arm. Newton, as if realizing the significance, let out an approving meow that echoed in the silence.

Leo finally found his voice, which came out somewhere between a squeak and a chuckle. "Are you serious? This isn't some elaborate prank, right? Like, Newton didn't hack into my phone and change voices, did he?"

There was a pause on the line, followed by a soft chuckle. "We assure you, Mr. Magnus, cats are yet to join the Nobel Committee. Congratulations to you both."

The Announcement, and a Bigger Surprise - The news spread like wildfire. Congratulatory emails flooded their inboxes, journalists clamoured for interviews, and colleagues from across the globe sent enthusiastic messages peppered with emojis of microscopes and beakers. Amid the chaos, Mara took Leo aside one evening, a secretive smile on her face.

"So, there's something I need to tell you," she said, her eyes twinkling with mischief.

Leo, still slightly dazed from the Nobel news, tilted his head. "Unless you're about to say we've also invented time travel, I don't think I can handle more surprises."

"Well," Mara said, taking his hand and placing it on her stomach, "this one's a bit more down-to-earth."

It took Leo a moment, but when realization struck, his eyes widened. "We're having a baby?" His voice cracked with pure joy.

Mara nodded, and Newton, as if on cue, leapt up and purred around their legs. "And before you ask," Mara added with a grin, "no, we're not naming it after Newton."

A Return to Roots - With the Nobel Prize ceremony looming, Leo and Mara made a big decision: they would move back to Leo's hometown, where it all began. The announcement brought a wave of excitement among friends and family, who eagerly awaited their return.

The trip to Stockholm was a whirlwind. The grand hall gleamed with chandeliers that cast light on the assembled luminaries from every corner of the scientific world. Leo stepped onto the stage, adjusting his tie nervously. Beside him, Mara looked radiant, with Newton's antics already immortalized in the ceremony brochure as the "cat that pioneered."

"Ladies and gentlemen," Leo began, glancing down at his prepared notes, then folding them and placing them in his pocket, "I had a speech, but Newton sat on it this morning. So here we go off-script."

The audience chuckled, and Leo felt his nerves settle. He spoke about growing up in his hometown, tinkering with magnets in his room, and the many kitchen science experiments that ended in flour explosions and smoke alarms.

"And through it all," he said, casting a warm smile at Mara, "I learned that science isn't just a solitary endeavour. It's shared joy, laughter, and the occasional argument over which part of a project our cat had technically contributed to."[131]

As he closed his speech, the audience erupted in applause. Mara wiped away a tear, and even Newton, who had miraculously managed to nap on stage during the speech, awoke to meow approvingly.

---

[131] The Nobel Prize in Chemistry recognizes groundbreaking work that has conferred the greatest benefit to humankind. The prize's roots trace back to Alfred Nobel's will, intending to encourage discoveries that improve our world.

Back Home, Full Circle - Returning to his hometown was like stepping back into a time capsule, only now Leo was carrying a Nobel Prize, a child on the way, and the love of his life by his side. The streets seemed smaller but filled with familiar faces, each one eager to share in the couple's joy.

One evening, Leo and Mara hosted a dinner for their closest friends and family. The table was laden with dishes Leo remembered from childhood, and the room was filled with laughter as stories were shared. Newton, wearing a tiny medal replica around his neck, prowled between chairs, accepting titbits of food.

As dessert was served, Leo's best friend Tom stood up, raising a glass. "To Leo and Mara," he said, his voice both proud and teasing, "the only people I know who can win a Nobel Prize and still somehow burn toast."

The room exploded in laughter, Leo joining in until tears streamed down his face. He glanced at Mara, who reached over and squeezed his hand, their Nobel medals catching the warm light. It wasn't just a symbol of achievement but of every laugh, failure, and moment of discovery they'd shared.

Final Reflections - As the evening wore on and the guests trickled out, Leo sat quietly, listening to the soft sounds of their home. He realized that his greatest discoveries were not just in the realm of science but in the people and moments that had enriched his journey. His life was filled with both the serious pursuit of knowledge and the humour that kept it alive.

Newton curled up on his lap, purring as if to echo Leo's thoughts. "We did pretty well, didn't we, buddy?" Leo whispered.

Extended Reflection: The Everlasting Journey - As the warm glow of the evening settled, Leo leaned back in his chair, the room still echoing with the cheerful chatter of guests. The quiet hum of conversations fading into the distance made the house feel alive, like it was breathing after a long-held sigh. Newton had made his rounds, collecting an impressive array of nibbles and now curled contentedly in a soft patch of moonlight.

Leo gazed at the room, catching the last notes of laughter as Tom tried to explain — rather poorly — how he once fixed a toaster with duct tape and a paper clip. The story had gotten more elaborate over the years, now including a heroic chase with a rogue bread slice that none of them believed, but all of them cherished hearing.

A Walk Down Memory Lane - After the last guest had departed, leaving behind an air scented with cinnamon and memories, Leo and Mara sat on the front porch, gazing up at the stars. Leo wrapped an arm around Mara's shoulder, feeling the pulse of the town he loved, its streets filled with familiar echoes of school bells, cheering crowds at soccer matches, and the soft rustle of leaves from the woods he used to explore as a child.

"You know," Mara began, breaking the comfortable silence, "this house doesn't have any lab equipment. Are you going to be okay with that?"

Leo smirked, tilting his head back. "Give me a week. There'll be a microscope in the kitchen before you know it. And Newton will probably insist on having his own workstation."

At the sound of his name, Newton perked up, whiskers twitching. He stretched lazily before climbing onto Leo's lap, purring as if he'd just solved a complex theorem.

"See?" Leo said, scratching behind Newton's ears. "He agrees."

They sat there, the three of them, content to let the night soak in, each star above them a potential story waiting to be discovered.

The Small Lab That Could - Despite his claim, Leo managed to keep the kitchen microscope-free for an impressive three days. But soon, Mara found herself sipping coffee next to an oddly placed centrifuge. Leo defended it, saying it was merely "temporary until Newton figures out how it works." She raised an eyebrow, watching the cat walk across the kitchen counter, inspect the device, and knock over a salt shaker as if in response.

Newton's antics became a source of endless amusement. One morning, Mara found him staring intently at a pile of silica gel packets as if trying to deduce the secrets of the universe. Leo, ever supportive of his feline companion's supposed genius, whispered, "He's close to a breakthrough, I can feel it."

Mara burst out laughing. "I'm sure he'll crack the code of 'Do Not Eat' one of these days."[132]

A Nobel Speech for the Ages - The next phase of their life together came with speaking engagements and public events. Leo's fame had grown beyond the quiet halls of research labs to television talk shows where producers learned quickly not to ask him questions involving the phrase "in simple terms." Leo always obliged, but not without a twinkle in his eye.

"Dr. Magnus, how would you explain the importance of nanobot-embedded self-healing materials in just one sentence?" a host once asked with a nervous smile.

---

[132] Silica gel is a desiccant, meaning it helps absorb moisture. It's commonly used to keep products dry and safe from humidity, which can damage electronics and foods.

Leo leaned forward, deadpan. "Imagine if your toast could repair itself after being burnt by Tom's toaster."

The audience erupted into laughter, and Leo sat back, satisfied. Mara watched from the side of the stage, shaking her head. It was impossible to escape Tom's legendary toast.

Newton, who had secured a regular spot in interviews as "co-researcher," sat beside Leo, donning a tiny bow tie and giving the occasional, disinterested lick of his paw. He had become the face of relatable science, complete with memes that read, "Newton's Third Law: For every action, there's a cat plotting an equal and opposite reaction."

The Town's Most Unexpected Nobel Celebration - Leo's return home had sparked a wave of nostalgia in the town. Old friends gathered, some of whom still joked about the time he accidentally filled the science lab with purple smoke or when Newton briefly disappeared into a crawl space only to emerge covered in glitter.

The local school invited Leo to give a talk, where he stood in front of wide-eyed kids who had doodled equations and cats on their notepads. He told them stories about how experiments often went wrong before they went right and how important it was to keep asking questions, even when the answer was a cat paw batting at your notes.

"And remember," Leo said, concluding his talk, "sometimes the greatest breakthroughs happen when you're just trying not to set the fire alarm off."

A boy in the back raised his hand. "Mr. Magnus, did you ever get in trouble with the principal for that?"

Leo laughed, Mara joining in from the corner. "More times than I can count. But I like to think even the principal learned something about combustion that day."

Full Circle with Family - The year rolled by, bringing with it new traditions. Dinner tables became filled with the sound of forks clinking and laughter erupting over inside jokes that had aged better than fine wine. Newton now had his dedicated seat, complete with a placemat emblazoned with "Chief Feline Investigator."

One evening, as the whole family gathered for an impromptu celebration — the reason this time being Newton's fourth "lab-iversary" — Leo stood up, glass in hand.

"To the experiments that failed spectacularly, to the inventions that made no sense, and to the people and cats who made the journey worth every unexpected twist."

Tom, always quick with a joke, raised his glass. "And to toast that doesn't set itself on fire!"

The room burst into laughter, tears glistening in the eyes of friends who had witnessed Leo's life come full circle — from the kid who wore an oversized lab coat to the Nobel laureate who still found humour in the chaos of science.

Leo looked at Mara, who was holding their newborn daughter, her tiny fingers curled around a miniature beaker toy. Newton leapt from his chair and curled up beside the baby, purring contentedly.

And in that moment, surrounded by love, laughter, and the infinite potential of discovery, Leo knew this was not just an end but the start of countless new adventures waiting to unfold — each one sprinkled with *science, serendipity, and a whole lot of cat fur.*

THE END